Applied Hydrodynamics in Petroleum Exploration

Second Edition

Eric C. Dahlberg

Applied Hydrodynamics in Petroleum Exploration

Second Edition

With 230 Illustrations

Springer-Verlag
New York Berlin Heidelberg London Paris
Tokyo Hong Kong Barcelona Budapest

Eric C. Dahlberg
ECD Geological Specialists Ltd.
405 Cliffe Ave SW
Calgary, Alberta T2S 0Z3
Canada

Library of Congress Cataloging-in-Publication Data
Dahlberg, Eric Charles, 1934–
  Applied hydrodynamics in petroleum exploration / Eric C. Dahlberg.-2nd ed.
  p.   cm.
  Includes bibliographical references and index.
  ISBN 0-387-97880-1
  1. Petroleum—Prospecting. 2. Fluid dynamics. I. Title.
TN271.P4D275 1994
622'.1828—dc20   93-50624

Printed on acid-free paper.

Production coordinated by Chernow Editorial Services Inc., and managed by Henry Krell;
  manufacturing supervised by Vincent Scelta.
Typeset by Best-set Typesetter Ltd., Hong Kong.
Printed and bound by Edwards Brothers, Inc., Ann Arbor, MI.
Printed in the United States of America.

9 8 7 6 5 4 3 2 1

ISBN 0-387-97880-1 Springer-Verlag New York Berlin Heidelberg
ISBN 3-540-97880-1 Springer-Verlag Berlin Heidelberg New York

Preface

In the first edition of this book, we observed that it had been created to fill a need for a usable "self-contained volume on hydrodynamics" (and hydrogeology) that was written specifically for the petroleum industry, but could also serve the earth science community in general.

When the first edition was published (1982), M.K. Hubbert, the father of petroleum hydrodynamics, was approaching the final stages of his very productive career. For this reason, the book served as a vehicle to amplify his concepts and spread and stimulate applications of some of his theories and methods throughout the exploration sectors of the petroleum industry. This was accomplished by blending discussions of Hubbert's concepts with some of the procedures used by industry specialists to answer practical oil and gas questions. The simple aim of the book was to bring this material to the fingertips of working geologists and geophysicists, who were "evaluating the hydrocarbon possibilities in larger exploration regions or assessing the potential of small, local subsurface oil and gas prospects." It was also hoped that by treating areas of conceptual overlap between petroleum geology and ground water hydrology, workers in both disciplines would be brought into closer contact, resulting in mutual benefits gained through healthy scientific and technical interaction.

This remains our objective in the second edition, although it has become apparent that additional material is needed to satisfactorily achieve it. The size of this volume reflects the new subject matter. Ten years of teaching in this subject have vividly exposed both the shortcomings and the accomplishments of the first edition. In this edition, we have sought to correct some of the former with new subject matter, while attempting to sharpen the focus by redoing our treatment of much of the original material.

The concern for hydrodynamics (more appropriately called petroleum hydrogeology) among petroleum geologists has increased significantly in the last few years for several reasons. First, limited employment opportunities in oil and gas have made it necessary for many of us to broaden our professional scope to facilitate career moves into environmental fields as hydrogeologists. These moves are facilitated by familiarity with subject matter common to both disciplines, such as hydraulic head, Henri Darcy's basic experiments, and inference of subsurface water-flow patterns, to mention a few.

Second, more effective formation-pressure measuring equipment has been developed, such as the amazing repeat formation tester (RFT). As a result, much more accurate subsurface data are now available for analysis and meaningful interpretation.

Third, there has been a great proliferation in the last ten years of computers and widespread availability of extremely powerful modeling programs, which can mimic the effects of underground hydraulic geologic processes in mere seconds.

Fourth, there has developed in the industry a tendency among individuals to examine more closely some of the less traditional exploration tools, such as petroleum hydrogeology. There are certain subsurface questions that can be answered using the procedures covered in this book. These include questions on predicting positions of hydrocarbon–water contacts that have not actually been penetrated by the drill (critical in estimating reserves); discerning reservoir trends using pressure–depth–gradient relationships (this principle triggered the discovery of one of the largest gas fields in western Canada by Canadian Hunter a few years ago) and testing geologically determined structures to predict whether the hydrologic and geologic components combine favorably to produce entrapment, or whether their interaction is destructive to hydrocarbon accumulation. Much of this methodology is appropriate to evaluating subsurface units for purposes of waste storage and disposal, as well as to exploration problems.

Much interest in abnormally pressured geologic units that are frequently encountered in drilling operations now exists, and it is apparent that the same hydraulic principles are relevant to fluid-related abnormal pressures as hydrodynamic phenomena.

Added material includes the following:

Improved figures to more effectively illustrate important concepts and relationships.

Discussion of drillstem testing, pressure chart extrapolation, repeat formation testing equipment and measurements, and other sources of formation pressure data, such as mud weights.

Discussion of the mechanics of abnormal pressure development and the geologic scenarios under which they occur, including osmosis, abnormal and normal compaction, fossil pressures, and the effect on supported and nonsupported seals.

An extensive Reference section, including relevant titles in the fields of hydrodynamics and geopressuring.

Whereas the first edition of this book was dedicated to the late Dr. J.C. Griffiths, our graduate school mentor, we are dedicating the present volume to a group of professional colleagues and personal friends. These are Roger L. Ames and Charles Bartberger of AMOCO; Sebastion Bell of the Geologic Survey of Canada; John S. Bradley, recently retired from AMOCO; Graham Davies and Bill Kerr, both Calgary consultants; David Matuszak of AMOCO; Hugh Reid of Hugh Reid and Associates; Wayne Rutledge, a Calgary consultant; and Herbert Sullivan, recently retired from AMOCO.

We hope that the reader will profit significantly from our present treatment of the still-fascinating subject of applied hydrodynamics in petroleum exploration, and that additional discoveries, as well as increased recoveries from already discovered pools, will ensue.

Eric C. Dahlberg

Contents

1

Fluids, Pressures, and Gradients

The Nature of Fluids

To facilitate the following treatment of the subject of applied hydro-dynamics in petroleum exploration and formation pressure geology, it is necessary to review some of the basic attributes of fluids. We will then be able to discuss the way fluids respond to external and internal influences in different subsurface geohydrologic environments.

The most important property affecting subsurface fluid behavior is *density*. The density of any substance is the relative weight of a given volume of it, such as X grams per cubic centimeter or Y pounds per cubic foot (e.g., a cubic foot volume weighs Y pounds). Oil densities are often expressed in kilograms per cubic meter (kg/m^3), with conventional crude oil densities ranging from $780\,kg/m^3$ to $900\,kg/m^3$ with most commonly in the middle 800s. Heavy oils, by definition, are those with densities exceeding $900\,kg/m^3$. Gas densities, relative to air at one atmosphere (1 atm), range between $0.55\,kg/m^3$ and $0.95\,kg/m^3$, although they are also reported relative to fresh water.

To eliminate the inconvenience of working with the different systems of units, we use the *specific gravity* of fluids, which is the density in appropriate units, compared with that of pure water in the same units (1.0000 gm/cc or $1,000\,kg/m^3$). Numerically, this is just the density value in decimal form, with no units! Standard temperatures and pressures are assumed.

In the petroleum industry, fluid densities are commonly described on an *American Petroleum Institute (API) gravity* scale. The formula that relates API gravity in degrees is

$$\text{API gravity} = \frac{141,500}{\text{density } (kg/m^3)} - 131.5. \qquad [1.1]$$

(Note that the API gravity of pure water is 10 degrees.)

By consequence, API gravity is the opposite of specific gravity, so an oil sample described as a high-gravity crude is really a lightweight oil, and heavy oils are those with low API values.

1

Pressures and Fluids

Any body of fluid (e.g., a pitcher of water, a cup of coffee, a reservoir of oil) has, with respect to pressure, the following attributes:

1. The internal pressure increases with depth in the body.
2. The rate at which the pressure increases is called the *static pressure gradient*, and it depends only on the density of the particular fluid concerned.
3. The two- or three-dimensional orientation of the vector representing the direction of maximum rate of pressure increase is vertical if the fluid concerned is static (motionless).

Furthermore, the pressure–depth relationships are completely independent of the shape of the fluid's container, be it a whiskey bottle, pinnacle reef, or sandbar encased in impermeable shale. Pressure–depth relationships are illustrated in Figures 1.1 and 1.2. Figure 1.1 illustrates the pressure gradient relationships within a single standard container; Figure 1.2 shows the same relationships in several differently shaped vessels. The containers each constitute a single, isolated system.

The pressures at all points within a confined fluid body (or system) plot graphically on a single straight line which represents the pressure gradient. At any point on the line, the pressure is dependent on three factors:

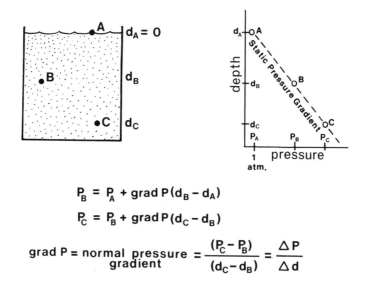

$$P_B = P_A + \text{grad } P(d_B - d_A)$$

$$P_C = P_B + \text{grad } P(d_C - d_B)$$

$$\text{grad } P = \text{normal pressure} = \frac{(P_C - P_B)}{(d_C - d_B)} = \frac{\Delta P}{\Delta d}$$

FIGURE 1.1. Pressure–density and gradient relationships in a static body of fluid.

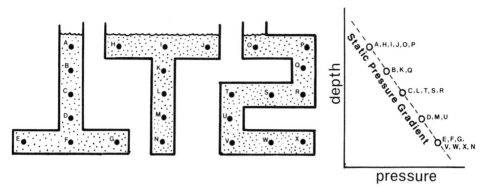

FIGURE 1.2. Container shape has no effect at all on hydrostatic gradient, which is the density-dependent rate at which the pressure increases with depth in the fluid body. Top of fluid column is 1 atm in all cases. Pressures from separate vessels all plot on the same gradient.

1. The density of the fluid itself
2. The depth of the point below the top of the fluid column
3. The pressure at the top of the fluid column

The pressure build-up with depth is attributable to the increasing weight of the fluid column above the particular point concerned and the rate at which the pressure increases downward with depth. (dP/dZ), the graphical slope of the pressure gradient, is numerically equal to D × g, where D is fluid density as mass per unit volume and g is the acceleration of gravity that converts the expression to weight (force) per unit volume. The pressure gradient (grad P = dP/dZ) can be calculated for practical purposes using the specific gravity and the following equation:

$$\text{grad P} = \frac{\text{specific gravity} \times 62.4}{144} = \frac{dP}{dZ} \qquad [1.2]$$

This relationship is illustrated schematically in Figure 1.3, which uses as examples a cubic foot of pure water, a cubic foot of oil, and a six-foot column of pure water. The force downward exerted by the one-foot cube of pure water is 62.4 pounds (force) (62.4 lbf). This is exerted over the area of the bottom of the cube, which is 144 in^2. Since pressure is essentially force per unit area, the pressure beneath the base of the one-foot cube of water is 62.4 lb ÷ 144 in^2, or 0.433 pounds per square inch per foot (psi/ft). For the stack of six cubic feet of pure water, the total force downward (its weight) is 6 × 62.4 = 374.1 lb. Converting this to pressure gives 374.1/144, or 2.59 psi. Thus, the pressure at the bottom is 2.59 psi greater than at the top, so the gradient is 2.59 psi over six feet of column, or 0.43 psi/ft, as before. A cubic foot of oil with a density of 795 kg/m^3 (80% as heavy as the water) will exert a force of 0.795 ×

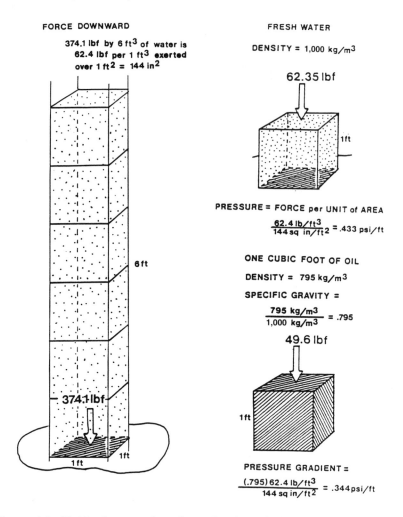

FIGURE 1.3. Fluid columns and gradients showing principles for calculating pressure gradients from density and specific gravity.

62.4 lbf = 49.6 lbf spread over 144 in² of area, so that the pressure gradient
becomes 49.6 lb/ft³ over 144 in², or 0.344 psi/ft. Thus, the above formula
compares any particular fluid with water and yields the gradient on a per-
foot basis accordingly.

In Figure 1.4, these characteristics are illustrated for six fluids commonly
dealt with by petroleum geologists. Each is contained within a hypothetical 1,000-ft-high closed vessel with a tight piston lid at the top of
each. This mechanism has been used to apply direct force to the contained
fluids, sufficient to elevate the pressure at the very top of each column to

FORCE TO ELEVATE PRESSURE AT TOP OF FLUID COLUMN TO 1,000 psi

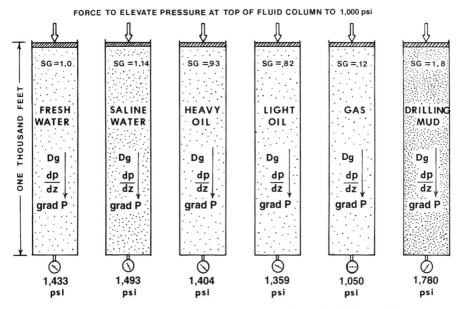

FIGURE 1.4. Examples of hydrostatic pressure gradients for different fluids.

1,000 psi. If the tanks were open, the pressures would be 1 atm rather than 1,000 psi, as a result of just the weight of the overlying air column. The pressures are indicated at the bottoms of the tanks, reflecting the height–weight–density differences.

Conversion of Units

In the oil industry, because some countries use the metric system and others do not, reported data often appear in units that have to be converted for pressure gradient plotting and other purposes. It is helpful to have the proper conversion factors at hand. These are presented in Table 1.1. For example, if one desires an estimate of formation pressure at a particular depth in pounds per square inch (psi), and the available data offer mud weight density in pounds per gallon (lb/gal), the steps for conversion are as follows:

1. Multiply lb/gal by 0.0519 (from table) to get the pressure gradient in psi/ft.
2. Multiply this gradient value by the required depth in feet to obtain the answer in psi.

As another example, if the lithostatic (overburden; (rock plus fluids)) gradient is 1.0 psi/ft (this is the commonly-accepted value), how much is it

TABLE 1.1. Conversion factors for gradients in metric and SI units.

To Get	multiply	by
kPa/m	psi/ft	22.5
kPa/m	lb/gal	1.169
psi/ft	kPa/m	0.044
psi/ft	lb/gal	0.0519
psi/m	psi/ft	3.28
psi/m	kPa/m	0.1450
lb/gal	psi/ft	19.3
lb/gal	kPa/m	0.852

in metric units (kPa/m)? The answer is 1.0 × (from the table) 22.5 = 22.5 kPa/m.

Table 1.2 is useful for determining fluid gradients. It offers the relationships among the different parameters discussed so far and can be used for comparing equivalent properties of a variety of subsurface fluids. For example, a brine sample with 145,000 parts per million (ppm) total dissolved solids would have a corresponding specific gravity of 1.1. At the same time, formation water from the Belly River formation (the hydrostatic pressure gradient for which is listed as 0.437) would obviously be very close to fresh, since the corresponding concentration of dissolved solids in the table would be around 10,000 ppm.

When chemical analyses of formation fluids are unavailable, estimates of water salinity can be made using resistivity log curve values; the lower the resistivity, the more saline the water concerned (and thus the higher its specific gravity and density). Deep resistivity values from well logs can be converted to sodium chloride concentration by calculation or by utilizing the graph in Figure 1.5, which provides the relationships between resistivity recorded in ohm-meters (ohm-m) and salinity in ppm as a function of temperature (which can be read from the appropriate well-log heading). For example, a formation water reflected by a log resistivity value R_w of 0.3 ohm-m at 140°F would have a corresponding salinity of 10,000 ppm (note the point plotted on the graph).

Gradients and pressure values at various points within static fluid bodies can be calculated using the simple equations previously presented in Figure 1.1. This figure illustrates a vessel of fluid with three points of reference at different depth locations, A at d_A, B at d_B, and C at d_C. The pressure at d_B relative to P_A is

$$P_B = P_A + \text{grad } P \, (d_B - d_A). \qquad [1.3]$$

In the same way,

$$P_C = P_B + \text{grad } P \, (d_C - d_B), \qquad [1.4]$$

TABLE 1.2. Relationships among specific gravity, API gravity, hydrostatic pressure gradient (psi/ft), and total dissolved solids for brines. Also, some typical hydrostatic pressure gradients for formation waters in some Alberta units.

Specific gravity	API gravity	Hydrostatic pressure gradient	Total solids (ppm)
2.5	−7.5°	1.083	210,000
2.0	−5.2°	0.866	175,800
1.5	−2.7°	0.650	143,500
1.25	3.0°	0.541	69,500
1.20	10.0°	0.520	0
1.14 (brines and heavy oils)	17.0°	0.494	
1.12	25.0°	0.485	
1.10	35.0°	0.476	
1.05	45.0°	0.455	
1.00 (fresh water)	60.0°	0.433	
0.95		0.411	
0.90		0.390	
0.85 (light oils)		0.368	
0.80		0.346	
0.70		0.303	
0.55		0.238	
0.50		0.216	
0.40 (gas)		0.130	
0.20		0.086	
0.15		0.065	
0.10		0.043	

$$\frac{\text{Specific gravity} \times 62.4}{144} = \text{grad P}$$

Water body	Hydrostatic pressure gradient
Belly River	0.437
Cardium	0.437
Milk River	0.440
Dunvegan	0.450
Blairmore	0.461
Leduc D-3	0.478
Beaverhill Lake–Swan Hills	0.474
Slave Point	0.474
Gilwood, Granite, Washington	0.474

Note relationships of gradient value to relative position in the geologic section. Variation reflects salinity differences.

where grad P is the appropriate static pressure gradient for the fluid concerned.

With two pressure values at diverse depths in any fluid system, the normal pressure gradient can be determined using the following formula:

$$\text{grad P} = \frac{(P_C - P_B)}{(d_C - d_B)} = \frac{\Delta P}{\Delta d}. \qquad [1.5]$$

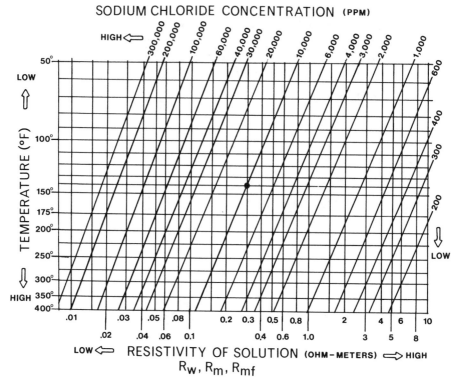

FIGURE 1.5. Chart for determining salinity (ppm NaCl) from log resistivity (ohm-m) and formation temperature (°F).

Ultimately, from the gradient, the density and the API gravity can be obtained for the fluid in question by working backwards through the proper formulas, described earlier.

Pascal's Principle

Illustrated in Figure 1.6 are the effects of Pascal's principle that are relevant to interpreting subsurface pressure pattern relationships. The principle states that when an external force is applied to a confined body of fluid, the internal pressure increases by the same amount at every point throughout the system. An enclosed body of fluid is shown capped by a lid that is completely supported by the walls of the container. The force exerted on the fluid (i.e., the weight of the lid) is therefore zero. The pressure at reference point A, focused from all directions, is proportional only to the weight of the fluid above it (as represented by the white arrows). Deeper in the container at point B, the pressure is even greater (as similarly represented by the white arrows). In the adjacent vessel, an external force is shown being applied to the confined fluid so that the pressure (which is force per unit area) at the top of the fluid column immediately below the lid is elevated by an amount proportional to the magnitude of the force, F. The pressure at point C is duly elevated (note the length of the white arrows), as it is at point D. The increase is proportional only to the force-related pressure increase.

When these four pressures are plotted on the accompanying pressure–depth graph, P_A and P_B lie (as they should) along the same static gradient,

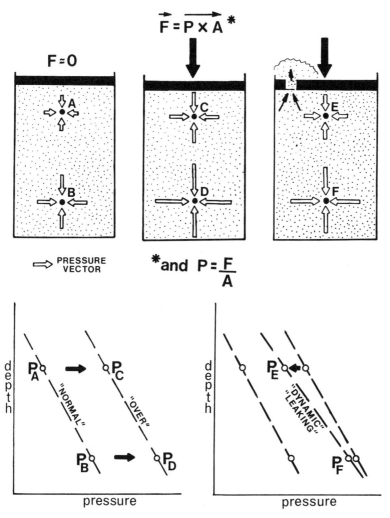

FIGURE 1.6. Mechanics and pressure–depth gradient relationships that demonstrate of Pascal's principle.

the slope of which depends only upon the density of the fluid. Points P_C and P_D, at identical depths to those of P_A and P_B, respectively, lie along a parallel gradient, which displaced toward higher pressures, reflects the "overpressured" condition created by the external force to which the amount of excess pressure is proportional. This example illustrates an important effect in the geologic interpretation of pressure–depth plot patterns.

   If, following the application of the external force to the body of confined fluid, the fluid became "unconfined" from a leakage, as shown in the third vessel in the figure, the resulting slow escape would be reflected in the pressure–depth relationships. The leaking system becomes dynamic as the fluids flow out. This change in the system is then reflected by the positions of points P_E and P_F on the pressure–depth plot. It appears that a gradient with a slope that doesn't correlate with the density of the particular fluid concerned thus reflects some leakage.

Pressure Relationships in Fluid Networks

Inside rocks, fluids are not present as a continuous unbroken phase, as in a container, but as a network within the pore spaces and throats between the mineral grains that make up the matrix framework of the rock material. In extreme cases, the formation water constitutes the thinnest of coatings adhering to the walls of the intergranular spaces, with the bulk of the space occupied by hydrocarbons, clays, or other substances. Thus, to be realistic, the pressure–density–depth relationships for water must be considered in the form illustrated schematically in Figure 1.7. This figure shows a fluid system in communication with the atmosphere at the two open necks, with some trapped gas in several of the individual chambers. Pressures from locations at vaious depths in the system are plotted on the graph at the right. All pressures in the fluid lie along the same line, which represents the normal static gradient for the system. The pressures in the confined gas portions of the system plot on the high side of the normal static gradient, since the gas is compressed by the water pressure (volume decrease with corresponding increase in internal pressure) to an amount proportional to the relative depth below the surface of the gas–water interface. For example, the pressure in gas body P,Q (which is the deepest in the system) is greater than that of chamber H, and both are greater than B, which itself exceeds atmospheric pressure (P_A) because of depth-related hydraulic compression.

   Figure 1.8 shows the system from Figure 1.7 with some changes. The original system has been rearranged into three separated systems, each of which is completely self-contained. System 1 is confined and exposed to heat, which raises the internal fluid pressure significantly. When the pressure reference points are plotted on the pressure–depth graph, the

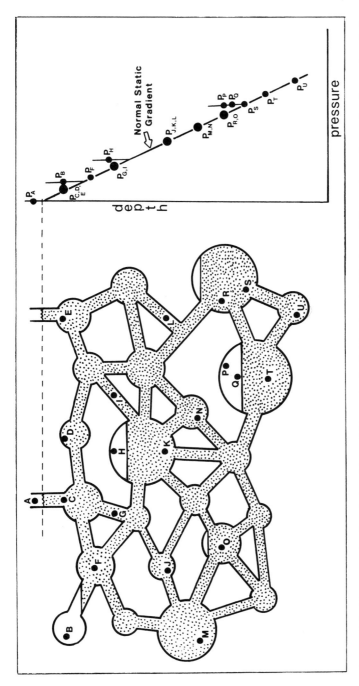

FIGURE 1.7.   Pressure–depth relationships within a single fluid network. The top of the water column for the system is in contact with the atmosphere. (Note chambers of trapped air included in the system.)

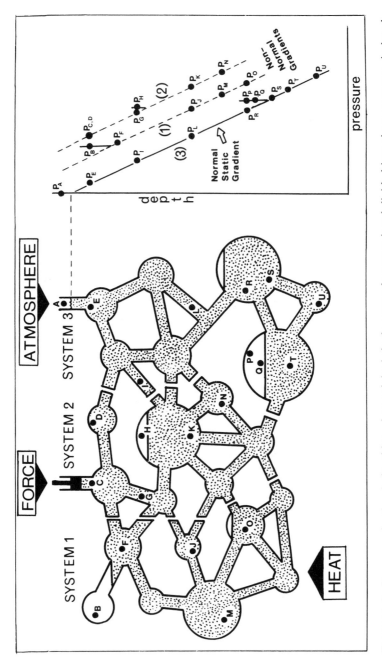

FIGURE 1.8. Pressure–depth relationships when original single network system is subdivided into three separate, isolated systems with the fluids acted on by different external influences.

line along which they lie reflects an elevated gradient for this system. The fluid in System 2, independent of the others, is under compression from the small piston located above chamber C, to which a force is being applied. The resulting proportional elevation in pressure at all locations within the system (Pascal's principle) is reflected in the gradient, which is displaced significantly to the right of the normal static gradient. The normal gradient reflects conditions in the unstressed network, System 3. Since it is an open system exposed to the atmosphere, System 3 functions as the normal standard with which the others are compared (and against which they are seen to be elevated). Pore fluids in rocks appear to exhibit the same relationships.

Planes and Gradients

In subsurface hydrodynamics, where the investigator cannot actually view directly the fluids of interest, the attributes of the hydrologic environment must be inferred from the construction and resulting orientation of

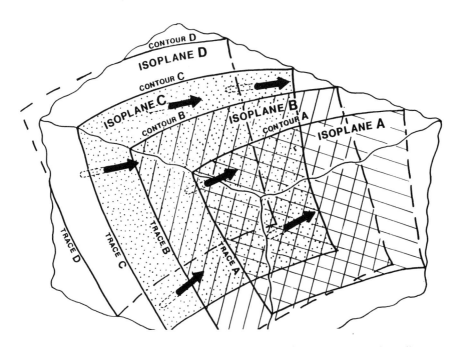

FIGURE 1.9. Three-dimensional relationships among isoplanes, normal gradients, vectors, and traces of the planes. These are required to specify subsurface fluid-movement directions and oil–water and gas–water contacts and to construct cross sections and maps.

hypothetical planes and gradients in three dimensions. Flow directions are then predicted relative to the resulting gradients. These represent paths along which hydrocarbons are most likely to travel in two or three dimensions. So, to predict the behavior of several different fluids, it is necessary to utilize geometric constructs, such as those illustrated in Figure 1.9. This figure shows a system of isoplanes and normal vectors in three-dimensional perspective. A family of planes such as these fix the gradients (black arrows) normal to them and increasing from D to A. Within each plane, the property concerned is at constant level, so that if these represent planes of constant salinity, level D would be the least salty, with increases in salinity up to level A. The black arrows represent the orientation of the actual salinity gradients increasing toward the right and out of the plane of the paper. The traces of the intersections of these planes of constant salinity with the top of the illustrated block are essentially contours representing levels of salinity, and the top of the block is a map. The traces along the side of the block represent the intersections of the isoplanes and the vertical edge plane. For example, if the orientation of the planes of constant potential energy for an oil occurring within the three-dimensional space of an aquifer or subsurface reservoir can be constructed like this, then the overall movement of any oil that finds its way into that space can be predicted as a function of the orientation of the internal forces which are normal to the planes.

2

Formation Pressure Measurements and Data

Sources of Pressure Data

Ideally, subsurface hydrodynamic procedures require most simply the best possible estimates of the virgin formation pressures at the particular depths within the geologic units concerned. Most of the available pressure data come from drill stem tests (DSTs), which have been routinely run by the industry for the past thirty years or so, and from repeat formation testing (RFT) tools, which have only recently begun to be commonly used in petroleum exploration.

Without being overly technical, Figure 2.1 illustrates the basic DST formation testing procedure. A tool consisting of the components illustrated in the figure is lowered through the mud column in the hole. This is the *running in* portion of the test, and while the *tester valve* is closed, the *ports* in the wall of the tool are open so that the drilling mud passes easily through the inner chamber of the tool as it travels down the hole. The pressure-recording gauges respond to the continuously changing fluid pressure in the mud column, which increases hydrostatically as the tool goes deeper and decreases when the tool is raised. When the formation to be sampled is reached (either at the bottom of the hole, as illustrated, or up at a shallower depth, in which case another *packer* is included so that the tool "straddles" the formation), the packer is expanded (donut-style) against the wall of the hole, as illustrated in (2). The tester valve is then opened, creating a pressure drop in the tool system (compared with the pore pressure within the formation). In response to this difference, fluids are sucked out of the rock into the tool and flow up past the pressure gauge. This is the *flowing period* of the test, during which a sample of the formation fluids (oil, water, gas, drilling mud, mud filtrate, etc.) is collected. In (3), the tester valve is closed, trapping the fluid sample in the chamber above it (this is the *shut-in period*); the pressure within the closed tool builds up, eventually reaching equilibrium with the pore fluid pressure of the isolated formation. These flowing and shut-in periods can

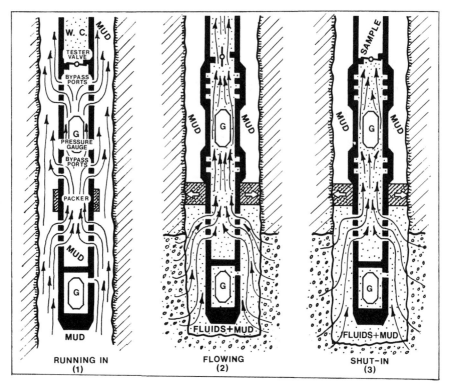

FIGURE 2.1. The episodes of a typical DST formation test.

be repeated several times for verification. At the conclusion of the test, the packer is collapsed and the tool is pulled out of the hole, set up as in (1) but now carrying a sample of the recovered formation fluids. The decreasing hydrostatic pressure of the mud from testing depth up toward the top of the hole is recorded on a chart as the tool rises to the rig floor.

DST Pressure Charts and Pressure Extrapolation

During the steps just described, a graph similar to that in Figure 2.2 is created, showing pressure variation over time. Note that the periods are labeled to correspond with the testing tool operations illustrated in Figure 2.1. Generally, the individual time periods vary in length; and the longer the period, the more reliable the resulting pressure reading, particularly in the case of the shut-in pressure, where the fluid pressure in the inner chamber of the tool builds toward equilibrium with the formation pore pressure which is always greater.

Since the objective of formation testing is to obtain an accurate value for the original formation pressure, allowing enough time for pressure buildup during the shut-in period is a necessity. Often, however, the test is terminated before the true, maximum formation pressure is reached. Since the buildup curve follows an essentially logarithmic path, a graphical (mathematical) procedure was devised to predict what the pressure curve would have built up to if the shut-in period had been infinitely long. This is done by extrapolation following these steps:

1. The existing buildup curve is sampled over a convenient number of time intervals (say, ten pressure values).
2. Δ-t (sometimes seen as θ) is defined as the elapsed (or cumulative) time in minutes from the beginning of the shut-in period. For example, the Δ-t values for a pressure curve sampled at five-minute intervals would be 0-, 5-, 10-, 15-, 20-, 25-, . . . , minute points.
3. The duration of the flow period is designated as T (see Figure 2.2), and with it the following simple parameters are calculated:

$$T + \Delta t \qquad\qquad [2.1]$$

and then

$$\frac{T + \Delta\text{-}t}{\Delta\text{-}t}. \qquad\qquad [2.2]$$

Obviously, if the tool were shut in for an infinite period, the elapsed time, Δ-t, would approach infinity (its limit). As this occurs, T + Δ-t (the

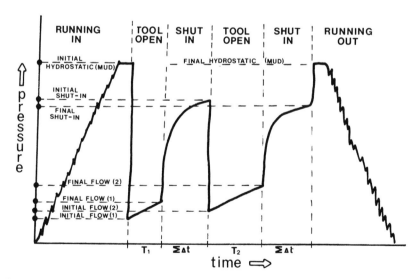

FIGURE 2.2. A model DST pressure–time graph, showing variation in gauge pressure during the test and the reported hydrostatic, initial shut-in, (ESI) and final shut-in (FSI) values and the points on the curve that reflect them.

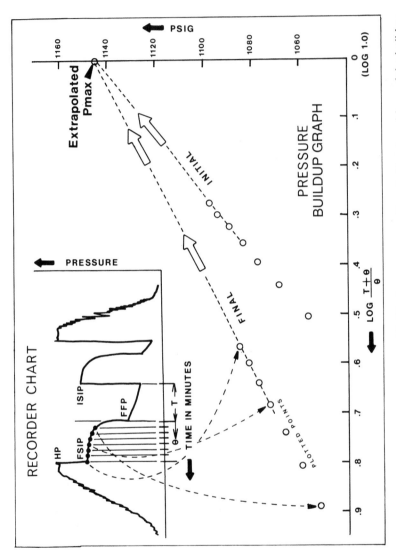

FIGURE 2.3. Example of an extrapolated DST curve constructed by plotting logarithms of the buildup ratio on standard equal-interval graph paper.

numerator) becomes huge, and T, compared with Δ-t, becomes miniscule and vanishes, so that the ratio becomes infinity/infinity, which is unity. Often, the greek letter θ (theta) is used in place of Δt in this expression.

4. On an appropriate graph, the individual pressure values from the original buildup curve (one for each each $T + θ$ value) are plotted against the logarithm of $(T + θ)/θ$ to linearize the curve. This can be done in two ways:
   a. The log values of $(T + θ)/θ$ can be plotted on equal-interval arithmetic graph paper, as shown in Figure 2.3.
   b. The raw values of $(T + θ)/θ$ can be plotted on semi-log graph paper with pressure along the arithmetic (vertical) axis, as illustrated in Figure 2.4; (the reader should complete the graph itself for practice).
5. The straight-line portion of the resulting points (this is often the last few points only) is projected linearly out to the appropriate pressure value corresponding to a limit of zero if the arithmetic plot is utilized (since log of 1.0 is zero) or a value of 1.0 on the logarithmic plot.

Δt (min.)	T + Δt (min.)	$\dfrac{T + Δt}{Δt}$	P_i (psi)
6			2759
12			2806
18			2827
24			2849
30			2859

T = 60

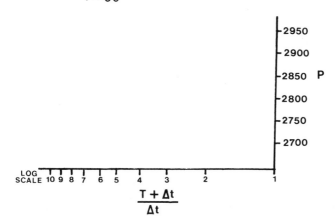

FIGURE 2.4. Sample pressure data for extrapolation using the original chart-selected buildup curve incremental values and semi-log graph paper. P_{max} is >2950 psi.

Figure 2.5 is a copy of an actual recorder chart from Haun (1982). The shapes of the curves (particularly the heavy, dark pressure buildup recorded as the testing tool is being lowered through the mud column in the well) disclose that the time origin is on the right end of the horizontal axis. The pressures corresponding to the curve can be seen in the pressure data portion of the table in Figure 2.6. The pressure values at five-minute increments are summarized in the lower half of the table, which contains all of the parameters required for extrapolation, using either of the graphing methods described earlier. Information such as instrument depth (5,874 ft), well temperature (140°F), and time length of the different periods is also included. Note that none of the three shut-in periods in this test reached the static reservoir pressure.

Accuracy of Drill Stem Test Measurements

The accuracy of drill stem test pressures is often discussed, but there appears to be a lack of definitive, statistically derived figures. Some that have been proposed are offered in Figure 2.7. Of these, number 4 seems

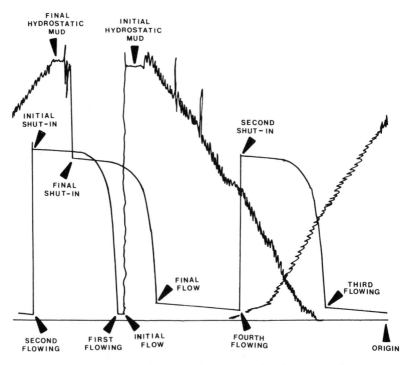

FIGURE 2.5. An actual recorder chart for a DST showing three shut-in periods superimposed over each other. (From Haun, 1982.)

This Chart From Haun (1982)

JOHNSTON TESTERS

PRESSURE DATA

rument No.	J-217			
Capacity (P.S.I.G.)	4700			Field Report No. __02987 B__
Instrument Depth	5874'			
Instrument Opening	OUTSIDE			
Pressure Gradient P.S.I./Ft.				
Well Temperature °F.	140			TIME DATA

					Time Given	Time Computed
Initial Hydrostatic Mud	A		3873			
Initial Shut-in	B	*	3657		60 Mins.	62 Mins.
Initial Flow	C		97		3 Mins.	3 Mins.
SECOND FLOW	C-4		924		30 Mins.	27 Mins.
SECOND SHUT-IN	B-1	*	3579		90 Mins.	93 Mins.
Final Flow	D		835		60 Mins.	58 Mins.
Final Shut-in	E	*	3574		180 Mins.	180 Mins.
Final Hydrostatic Mud	F		3842			
Remarks:	C-1		904			
	C-2		945			
	C-3		973			
	C-5		967			
	C-6		964			

*Shut in pressure did not reach static reservoir pressure.   Clock Travel __0.02031__   inches per min.

PRESSURE INCREMENTS

INITIAL SHUT-IN

Point	Pressure	$\frac{T+\Delta t}{\Delta t}$
C-1 0	904	
5	3112	1.600
10	3418	1.300
15	3515	1.200
20	3562	1.150
25	3591	1.120
30	3611	1.100
35	3624	1.086
40	3634	1.075
45	3641	1.067
50	3646	1.060
55	3651	1.055
60	3656	1.050
B 62	3657	1.048

SECOND SHUT-IN

Point Minutes	Pressure	$\frac{T+\Delta t}{\Delta t}$
C-4 0	924	
5	2498	7.000
10	2910	4.000
15	3111	3.000
20	3229	2.500
25	3309	2.200
30	3368	2.000
35	3409	1.857
40	3442	1.750
45	3468	1.667
50	3490	1.600
55	3508	1.545
60	3521	1.500
65	3535	1.462
70	3546	1.429
75	3554	1.400
80	3563	1.375
85	3570	1.353
90	3576	1.333
B-1 93	3579	1.323

FINAL SHUT-IN

Point Minutes	Pressure	$\frac{T+\Delta t}{\Delta t}$
D 0	835	
10	2686	9.800
20	3029	5.400
30	3192	3.933
40	3290	3.200
50	3355	2.760
60	3400	2.467
70	3433	2.257
80	3460	2.100
90	3482	1.978
100	3500	1.880
110	3515	1.800
120	3529	1.733
130	3538	1.677
140	3548	1.629
150	3557	1.587
160	3564	1.550
170	3569	1.518
E 180	3574	1.489

FIGURE 2.6. Example of a DST report, showing the recorded pressure data for three shut-in periods. These data can be plotted on blank arithmetic and logarithmic graph sheets. (From Haun, 1982.)

the most useful, but assessing the reliability of any subsurface pressure measurement is more of an art than a science, since so many uncontrollable factors and unverifiable effects can interfere with the measurement procedure. In most cases, however, one won't place much credence in pressure differences that do not significantly exceed the levels of accuracy suggested in this list.

1. **Roach (1965)**   ± 0.5 %   (1800 ± 9 psi, 2375 ± 12 psi)

2. **Lynes** (1975)   0.5 % of the total   (± 25 psi,
range of the pressure   ± 50 psi)
recorder.

3. **Bradley (1975)**   ±1.0 % of the full   (±50 psi,
gauge scale.   ±100 psi)

4. **Unknown**   ±1.0 % per thousand   ±35 psi at
Source   feet of depth.   3500 ft.

**In general, accuracy depends on magnitude of
the pressures measured.**

FIGURE 2.7. Several different published values reflecting the accuracy of DST pressure estimates.

Formation Pressure Information from Well Records

Well data cards and scout tickets (which are kept on file in most companies) are a useful secondary source of formation pressure information for the subsurface exploration geologist. While listing major formational tops and markers, they also give intervals that have been tested, the pressures measured during the various episodes in the test, and the types of fluids recovered. The geologist engaged in hydrogeologic work is concerned with collecting the most accurate estimate possible of the fluid pressure within the interval of interest. This will be one of the shut-in pressures reported as SIP (shut-in pressure), ISIP (initial shut-in pressure), and/or FSIP (final shut-in pressure). The nature of the fluid in which the pressure measurement has been made is likewise essential, since, for regional mapping, the pressure in a water-saturated part of the aquifer is most suitable. For reservoir continuity and fluid system correlations, oil and gas pressures are utilized, too.

In each of these cases, the initial shut-in pressure appears to be the highest and would thus be preferred as the basis for subsequent calculations. However, it isn't always this way; often, the final shut-in pressure is the highest and would then be used as the best estimate of the original, virgin pressure of the formation. On some occasions, neither of the recorded shut-in pressures is high enough (perhaps because of something like a tool malfunction or clerical error) so that none of the data can be applied to hydrodynamic calculations. It is a favorable indication, in terms of accuracy, when all the shut-in pressures in the same unit and well, actual or extrapolated, are in close agreement.

When the initial and the final shut-in pressures are significantly close to each other, such as in the third test (3,275–3,250), it is reasonable to assume that the highest of the two is the best estimate of the virgin, predrilling, natural pressure of the formation concerned. When the final shut-in pressure is considerably higher than the initial shut-in pressure, to an extent that it closely approaches the mud hydrostatic pressure (HP), it is likely that one of the packers has collapsed during the test and that the recorded pressure does not accurately reflect the formation pressure. The differences between the flowing pressures and the shut-in pressures furthermore reflect the permeability of the formation itself. When both are relatively high and close to each other, the formation is one within which the fluids probably move with ease. On the other hand, when the shut-in pressure levels are high but the flowing pressures are extremely low, the formation fluids are not moving easily.

Computer Data Bases

Some geologists rarely get to see an original drill stem test chart; they purchase their pressure-related data from venders who keep computerized drill stem records. Examples of different formats are illustrated next. The geologist with access to a spreadsheet program such as *Lotus 123* can then build a project data base according to a format that for example might include columns of test information such as well name and location, DST number, date, KB (elevation) of the well, top and bottom of the tested interval, formations within the interval, hydrostatic pressures, flowing pressures, shut-in pressures, elevation of the recorder, fluid recoveries, (P/D) ratio, and so on. Once these data are moved into the spreadsheet environment, calculations of parameters such as hydraulic head can be executed using cell functions and can be graphed using the appropriate associated software routines. Figure 2.8 is a reproduction of several drill stem test summary reports generated from the International Petrodata System.

In Figure 2.9, formation pressures from three sources are listed for purposes of comparison and to assess the degree of improvement provided by pressure curve extrapolation.

In the column under "PIX", the shut-in pressures reported on scout tickets for this group of wells are listed. These can be used in project work if they are the only data that are available. Listed under "Petrodata" are the same test results inferred from inspection of the original charts and offered to the public is an edited and cleaned up (but not extrapolated) form, i.e. computer data base. Of these, the highest is generally the most usable. These are underlined.

The fourth column contains the extrapolated results as "PMax" values. However, the last column shows changes relative to the highest unex-

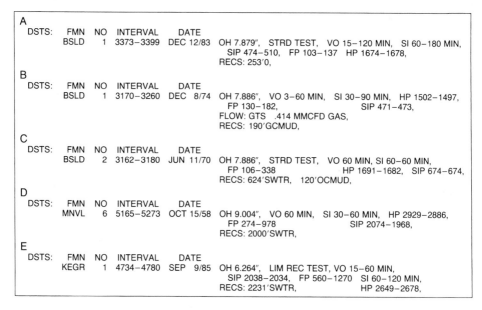

A
DSTS: FMN   NO   INTERVAL   DATE
        BSLD   1   3373–3399   DEC 12/83   OH 7.879″,   STRD TEST,   VO 15–120 MIN,   SI 60–180 MIN,
                                             SIP 474–510,   FP 103–137   HP 1674–1678,
                                             RECS: 253′0,
B
DSTS: FMN   NO   INTERVAL   DATE
        BSLD   1   3170–3260   DEC 8/74   OH 7.886″,   VO 3–60 MIN,   SI 30–90 MIN,   HP 1502–1497,
                                             FP 130–182,   SIP 471–473,
                                             FLOW: GTS .414 MMCFD GAS,
                                             RECS: 190′GCMUD,
C
DSTS: FMN   NO   INTERVAL   DATE
        BSLD   2   3162–3180   JUN 11/70   OH 7.886″,   STRD TEST,   VO 60 MIN, SI 60–60 MIN,
                                             FP 106–338   HP 1691–1682,   SIP 674–674,
                                             RECS: 624′SWTR,   120′OCMUD,
D
DSTS: FMN   NO   INTERVAL   DATE
        MNVL   6   5165–5273   OCT 15/58   OH 9.004″,   VO 60 MIN,   SI 30–60 MIN,   HP 2929–2886,
                                             FP 274–978   SIP 2074–1968,
                                             RECS: 2000′SWTR,
E
DSTS: FMN   NO   INTERVAL   DATE
        KEGR   1   4734–4780   SEP 9/85   OH 6.264″,   LIM REC TEST, VO 15–60 MIN,
                                             SIP 2038–2034,   FP 560–1270 SI 60–120 MIN,
                                             RECS: 2231′SWTR,   HP 2649–2678,

FIGURE 2.8. Several computer-generated DST summary reports, listing pressures and recovered fluids.

trapolated value, the extrapolations change the estimated formation pressure from −48 psi up to 106 psi, but the average change is about 10 psi. This is hardly significant, but the reader is definitely advised to use extrapolated test data when it is easily available; it is generally more accurate (a closer approximation of the "real" pressure).

Repeat Formation Testing (RFT)

A recent development in subsurface formation testing is the availability of tools that can (on a single lowering into a well bore) take several formation fluid samples at different depths and obtain formation pressure measurements at numerous intervals up and down the hole. Each testing is essentially a mini drill stem test executed over a significantly shorter time period than the conventional single-trip DST. The DST is, of course, a mini well completion in its own right, but it suffices to say that the RFT puts excellent information in the hands of the petroleum geologist, since it provides a series of actually measured pressures over intervals anywhere in a given well. For individuals concerned with gradients reflecting fluid environment, gradients reflecting fluid types, and the locations of hydrocarbon–water interfaces (which on the basis of DST-measured pressures have to be projected), the RFT tool defines them clearly and

Well Location	PIX	Petrodata	Extrapotated (P_{max})	Change psi
15-04-1-15W4	879 – 879	none	893	+14
01-09-1-15W4	863 – 817	877 – 821	877	0
02-09-1-15W4	900 – 835	884 – 832	912	+12
08-09-1-15W4	781 – 739	816 – 760	823	+7
05-10-1-15W4	851 – 822	869	905	+36
05-10-1-15W4	783 – 788	796	796	0
14-12-1-15W4	882 – 850	887 – 844	887	0
11-14-1-15W4	904 – 817	904 – 817	923	+14
04-18-1-15W4	909 – 909	954 – 954	954	0
03-02-1-16W4	932 – 773	937 – 775	889	−48
04-02-1-16W4	86 – 816	890 – 810	none	
08-02-1-16W4	761 – 761	775 – 767	821	+46
10-02-1-16W4	920 – 903	428 – 503	916	−4
02-03-1-16W4	681 – 685	677	791	+106
08-03-1-16W4	786 – 805	800	805	0
12-04-1-16W4	646 – 646	673 – 662	709	+36
12-04-1-16W4	753 – 753	755 – 752	733	−22
10-05-1-16W4	777 – 772	777 – 772	777	0
14-07-1-16W4	837 – 837	837	842	+5
01-08-1-16W4	820 – 770	801 – 789	798	−3
04-08-1-16W4	782 – 787	none	787	0
13-10-1-16W4	848 – 838	862 – 836	862	0
10-12-1-16W4	859 – 861	859 – 861	864	+5
05-15-1-16W4	833 – 835	837 – 839	841	+2
15-15-1-16W4	610 – 690	688	718	+28
12-17-1-16W4	898 – 826	883 – 822	925	+27
12-17-1-16W4	835 – 826	836 – 813	836	0
			Average Change is	+10

FIGURE 2.9. Comparison of formation pressure estimates for a collection of western Canada wells from scout data (PIX), a commercial data bank (International Petrodata), and P_{max} values extrapolated from the individual buildup curves.

explicitly. In the past, expensive drill stem tests were aimed at potentially productive units and pressures measured in water-saturated reservoirs which are required for mapping subsurface flow and executing other hydrodynamics procedures resulted accidentally. The RFT can be run anywhere in the well where pressure data is required, and since over-pressured zones often occur in nonpetroliferous portions of the section, geopressured zones can be recognized at their stratigraphic onset with this tool. The number of individual pressure tests executed over an interval of interest is almost unlimited, although fluid samples can be retrieved from only two or three.

The Selective Formation Tester

An example of a repeat formation testing tool is the selective formation tester (SFT), which is part of a system operated by the Computalog Gearhart company. The following description of a repeat formation testing procedure is based on this tool, as described in the company's manual. This very generalized treatment is included here to familiarize the geologist with the basic principles of this type of testing. The reference diagrams resemble the real tool only superficially and merely illustrate the RFT procedure in a most general way.

Figure 2.10 illustrates the major components of the tool, which include several motors, a hydraulic system for operating the sampling equipment, several sampling chambers for storing recovered formation fluids, and electrical transducers for recording pressure changes during the individual tests. There are also electronic devices for transmitting pressure data to recorders at the surface. This figure shows the tool set-up as it is lowered into a mud-filled well bore. Note that there are several openings in the tool where the mud (at hydrostatic pressure) is in contact with components of the hydraulic system in order to maintain a balance between them. As the tool is lowered, both the *formation sampler* and the *seating pad* are in retracted position for ease of entry, and the well mud in the *mini sample chamber* is in direct contact with the internal *sample pressure transducer* so that the hydrostatic pressure of the mud is continually monitored. The transducers are essentially wire-wrapped metal cylinders, parts of which expand in response to filling by fluids under pressure, so that the electrical properties of the stretched wire wrapping, compared with those of an unstretched portion, are recorded and translated into appropriate pressure units (kPa, mPa, or psi). Both the hydrostatic pressure measured in the sample pressure transducer and the hydraulic pressure measured by the hydraulic transducer are transmitted to the recorder chart at the surface. Note that the pathways from the hole to the main sample chamber from the mini sample chamber are closed off by valves.

Once the tester arrives at the chosen depth (this is usually selected from gamma ray log curves), the tool is set up to measure flowing and shut-in pressures for the zone concerned and to obtain a sample of the formation fluids present. As illustrated in Figure 2.11, the hydraulic pump expands the seating pad against the wall of the hole on the side opposite the formation sampler, which is at the same time pushed outward against and pressed into the formation to be tested. The mini sample chamber is then operated by the *sampler motor*, which pulls the cylinder backward, enlarging the internal volume of the chamber by a limited, calibrated amount (from 1 to 20 cc). Formation fluids flow into the chamber (shown in black). The pressure change from hydrostatic, to flowing, to shut-in (when the vessel is completely filled and the pressure equal to that in the formation pores) is charted on the recorder for later interpretation. The

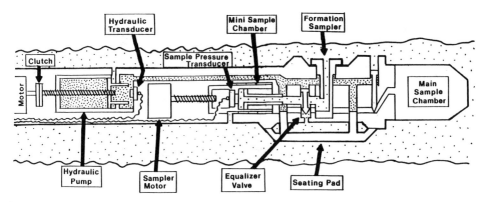

FIGURE 2.10. The components of the Computalog Gearhart selective formation testing tool as it enters and moves through the mud-filled well bore.

seal valve leading to the main sample chamber is opened by the elevated hydraulic pressure, but fluids are unable to pass into this chamber until access is granted from the sample valve at the mouth of the mini sample chamber.

As the sampler motor turns, the mini sample chamber recedes farther upward into the tool, eventually clearing the sample valve so that fluids can be conducted through the formation sampler past the seal valve and on into the sample chamber, as illustrated in Figure 2.12. The water cushion and the internal pistons and valves control these high-pressure fluids (oil, gas, formation water, or drilling mud). This sample is eventually expelled from that chamber with compressed air.

RFT Data

The heading of an RFT log from the Grand Banks run by Schlumberger of Canada is illustrated in Figure 2.13. As the label indicates, four separate intervals were sampled, and the tops and bottoms of these intervals are posted in the four "Run No." columns. (There are additional parameters on the heading as well, which may or may not be important to the user.) For the geologist, the next significant part of the log is the portion summarizing the results of the individual tests in each run (see Figure 2.14); these are found in the "TESTING DATA" section. Within each of these fields there are columns of data. The first column lists the "Test No." (from 1 to 9 for the first run and from 1 to 5 for the third). The second column records the "Depth" of the test to the nearest half meter (the depth for Test No. 1 of Run One is 3,369.5 m). The next four columns provide information on the sampler and the permeability of the formation in the vicinity of the test. (In the case of Run One, only

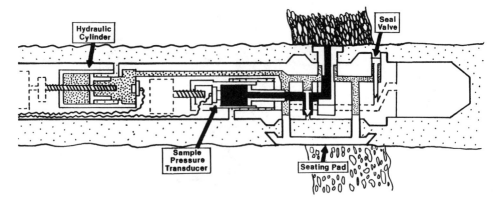

FIGURE 2.11. Operation of the SFT once the formation sampler is seated against the sampled unit. Fluids under pressure flow into the sample chamber for collection and pressure measurement.

hydrostatic [mud weight] pressures were recorded because of failure to seat the tool properly; in Run Three, initial shut-in pressures were recorded for three tests before the tool malfunctioned [these are listed in the seventh column].) The eighth and ninth columns indicate which tests provided formation fluid samples as well. In some cases, corrections for other effects, such as temperature, are made on the measured pressure values; this is duly noted on the log in the "REMARKS" section.

Figure 2.14 is actually the first portion of the log, which is the summary of the measurements for the individual runs described in the heading. Two are shown in this example. For each depth, there is a single numbered test and usually a pressure value (and some other information on sampl-

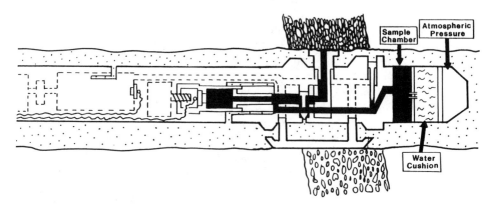

FIGURE 2.12. The SFT in the process of extracting a second sample of fluid from a formation.

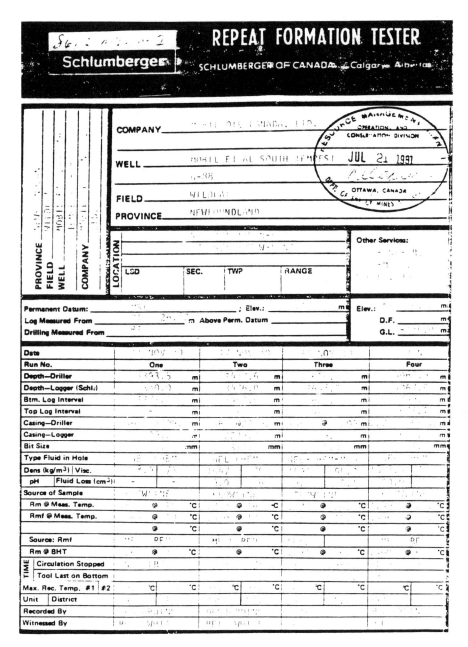

FIGURE 2.13. Example of a Schlumberger RFT log heading, showing the parameters recorded for each run and the number of runs.

Run No.	One						REMARKS		
Service Order No.	101947						RUN 1		
EQUIPMENT DATA							PRESS RETESTS		
RFP	AB	131					THE TESTS WITH NO SEAT		
RVP	AA	132					RECORDED.		
RPP	AB	133							
RFC	AA	280							
RFM	AB	91							
RFS									
RFS							Log Taped □Yes □No		

TESTING DATA

Test No.	Depth	Good Seat (✓) Yes	No	Permeable (✓) Yes	No	HYDRO-STATIC KPA	Sample Taken (✓) Yes	No	Test No.	Depth	Good Seat (✓) Yes	No	Permeable (✓) Yes	No	Initial Shut-in Pressure
1	3269.5					44850									
2	3266.5					44900									
3	3266.5					44850									
4	3331.5					44360									
5	3303.5					44090									
6	3369.7					44900									
7	3281.0					43840									
8	3256.5					43520									
9	3243.5					4355?									

Run No.	THREE						REMARKS		
Service Order No.	101949								
EQUIPMENT DATA							ENTER		
RFP	AA	132							
RVP	AB	132					NOTE DIFFERENCE		
RPP	AB	133					BETWEEN TESTS		
RFC	AA	280							
RFM	AB	91							
RFS									
RFS							Log Taped □Yes ⊠No		

TESTING DATA

Test No.	Depth	Good Seat (✓) Yes	No	Permeable (✓) Yes	No	Initial Shut-in Pressure	Sample Taken (✓) Yes	No	Test No.	Depth	Good Seat (✓) Yes	No	Permeable (✓) Yes	No	Initial Shut-in Pressure
1						49730									
2						49730									
3						49710									
4						51460									
5						51510									

FIGURE 2.14. Example of a portion of a Schlumberger RFT log showing the summarized pressure results for the individual runs.

ing, etc., if recorded). Run One includes nine individual tests, and Run Three lists five.

Figure 2.15 illustrates the recorded pressure curves on the RFT log. The scales for these traces are at the top of the trace; the analog pressure scale is general, covering the range 0–100,000 kPa over 2.5 in. The digital pressure traces are recorded logarithmically, with ranges of 0–100,000, 0–10,000, 0–1,000, and 0–100 kPa; the curve segments on each track must be added to yield the total pressure value concerned. In the example, the full shut-in pressure for the run is $50,000 + 1,000 + 500 + 10 = 51,510\,kPa$. What is the hydrostatic pressure? It is $54,220\,kPa$.

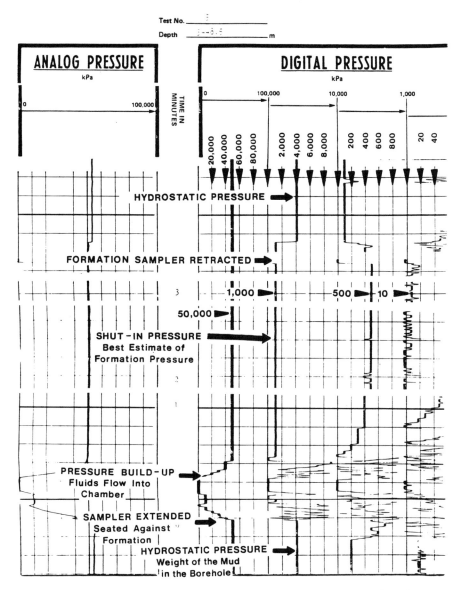

FIGURE 2.15. Example of the pressure curves for a single test from a Schlumberger RFT log. The events reflected by each curve are labeled, as are the scales. The individual curves resemble DST curves with analogous events.

Analysis and Interpretation of RFT Log Data

A set of pressure–depth data from one RFT run in the K-17 Terra Nova well has been input to a *Lotus 123* file titled GBKSRFTS.WK1 (for Grand Banks Repeat Formation Tests, Worksheet 1). In this run there are twenty-six tests done at various depths in the well between 1,334 m and 1,713 m; the shut-in pressures are in kilopascals. To interpret the data, the plot shown in Figure 2.16 has been generated using PGRAPH in *Lotus 123*. Each individual pressure value is represented by a square on the graph. Measurements from two different portions of the hole make up this single run: four at the 1,300-m level and the rest in the interval between 1,500 m and 1,700 m. Two distinctly different gradients

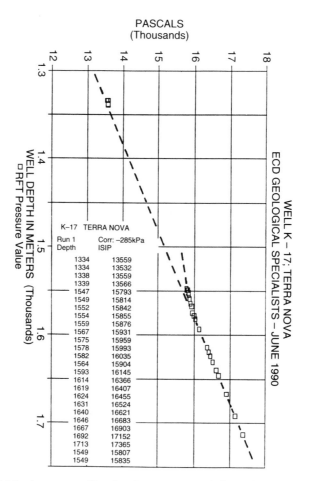

FIGURE 2.16. A pressure-Depth plot constructed from RFT data utilizing the Lotus 123 system.

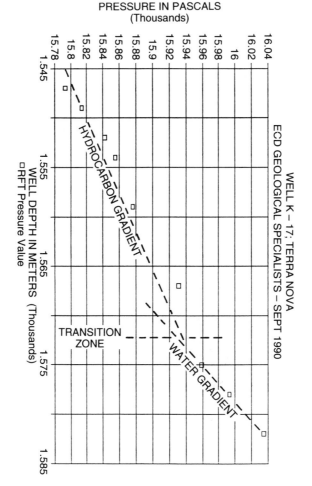

FIGURE 2.17. Enlarged gradient intersection portion of the RFT plot illustrated in Figure 2.16. Inferred transition zone is 1573 m.

are inferred from these point configurations, with a region of intersection indicating a fluid contact. Figure 2.17 is an enlargement of the interval between 1,545 m and 1,585 m, enabling the hydrocarbon–water contact to be more accurately placed.

The RFT and SFT tools have been a breakthrough in the field of petroleum hydrodynamics because they enable actual pressure measurements to be made over a geologic interval in a well, indicating the nature of the real pressure gradient. Prior to the development of these tools, such gradients could only be approximated by connecting widespread discrete points on a graph or calculating from fluid density assumptions.

3

Fluid Environments

The Hydrostatic Environment

A *hydrostatic* environment is one within which there is no internal motion or movement of the fluid. The maximum internal pressure gradient (the direction in which the rate of pressure increase is the greatest) is vertical, attributable to gravitational weight of the overlying fluids, as discussed earlier. All internal forces are oriented vertically, buoyancy being the major one. (Anyone who has swum to the bottom of a deep swimming pool has felt the effects of this gradient in the form of "squeeze.")

A mechanical "tank" model of a hydrostatic subsurface reservoir is illustrated in Figure 3.1. The figure shows the essential internal and external components and the dimensional variables from which the hydrologic parameters can be calculated. The model is a closed vessel of fluid with a piston-type lid to which an external force can be applied to raise the internal pressure within the tank. On one side of the main tank is attached a vertical manometer-type tube, which exits the wall of the chamber as illustrated. Obviously, the height of the fluid column in the tube is proportional to the pressure in the tank at the internal point of interest. It is also a function of the density of the fluid concerned and the acceleration of gravity (g). When a load is applied to the piston and transmitted to the top of the fluid body, the internal pressure is elevated. With knowledge of the density of the fluid (D), measurement of the elevation of the point of interest relative to a datum (Z), and measurement of the height of the water column in the manometer tube (H_W), the internal fluid pressure at point P can be calculated using the equations shown in the figure. Relevance of this model in a geologic setting is illustrated in Figure 3.2. The scenario shows a buried, carbonate reef the pore spaces of which are saturated with formation water of density D, under pressure P. In this case, hydraulic head is then calculated using the same relationships illustrated by the model, although the dimensions are in terms of well depths and sea level. Z becomes Kelly Bushing (KB) elevation of the drilling rig floor minus the depth of the pressure recorder (RD). For the tank model, H_W is actually measured and the relationships are used to cal-

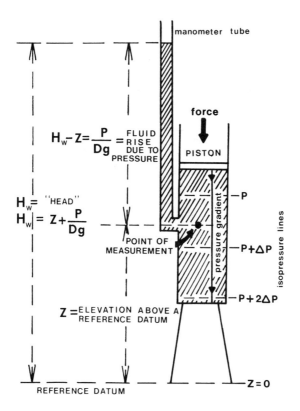

FIGURE 3.1. A "tank" model of a hydrostatic reservoir illustrating the relationships between internal fluid pressure at a point in the fluid body and the corresponding hydraulic "head" of the fluid at that point reflected by the height of the fluid column in a manometer tube.

culate the unknown pressure; in the subsurface application of the principle, the pressure itself is measured and the unknown hydraulic head then calculated to estimate the potential energy of the water itself for purposes to be described in subsequent chapters.

Internal Forces—Static Environment

To understand the behavior of three-phase oil, gas, and water systems, it is worth looking at force vector constructions that characterize hydrostatic and nonhydrostatic fluid environments. Figure 3.3 illustrates the combinations of forces that act on unit masses of water, oil, and gas in a hydrostatic system. The first major component is gravity directed vertically downward, as represented by vector g. The pull of gravity is constant for

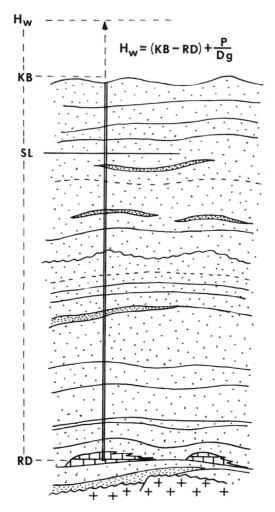

$$H_w = (KB - RD) + \frac{P}{Dg}$$

FIGURE 3.2. Relationships demonstrated in the tank model (Figure 3.1) shown in a geologic scenario using equivalent subsurface testing and well-logging parameters.

water, oil, and gas. In each case, the point is the center of the unit masses of water, oil, and gas, a unit mass being essentially that volume of the fluid concerned that as a function of its density will have a mass of exactly unity. The water, oil, and gas volumes occupied by the unit masses will be increasingly larger, since $m = VD$, $m = 1$, $1 = VD$, so $V = 1/D$.

Opposing the gravity vector is the buoyancy vector, which is a function of density differences and volume–pressure differential and reflects a force component oriented vertically upward. In physics this is symbolized as ρg or Dg. In the case of static water, these two forces are perfectly balanced so that the resultant force, represented in the figure as E_w, is

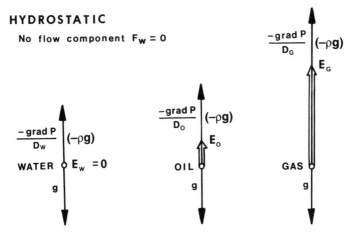

FIGURE 3.3. Vector model illustrating the principal forces that act on elements of water, oil, and gas in a hydrostatic environment.

zero. Thus, in a static body of water there are no force imbalances or motion, and level of potential energy of the water is constant throughout.

In the case of the oil, the buoyancy vector exceeds in magnitude the buoyancy vector for the water since the numerator D_O is less than D_W. The buoyancy force acting on the volume of oil located within the water, occupied by the unit mass $(1/D_O)$, is then $1/D_O$ times $(-\text{grad } P)$, the negative of the pressure gradient (negative because pressure decreases upward). Since this upward-oriented buoyancy vector for the oil exceeds the gravity vector oriented downward, there is a resultant force E_O, which is indicated by the white arrow.

The magnitude of the resultant force acting on the unit mass of gas contained in the water, represented by the white vector E_G, is the greatest of the three, since the gas density is the lowest of all, and the resulting difference between the buoyancy vector and the gravity vector is the greatest. Three Ping-Pong balls (one filled with water, two filled with hydrocarbons) being released from the bottom of a calm swimming pool would demonstrate the behaviors illustrated by the vector models. The water-filled ball would (assuming constant temperature, salinity, etc.) remain motionless at the release point, while the hydrocarbon-filled balls would rise vertically to the surface, with the least dense one moving the fastest and arriving the quickest.

The Hydrodynamic Environment

The *hydrodynamic* environment is one in which internal force imbalances do exist. The basic characteristics of the hydrodynamic environment are illustrated in Figure 3.4. This model is an elongated, closed vessel with

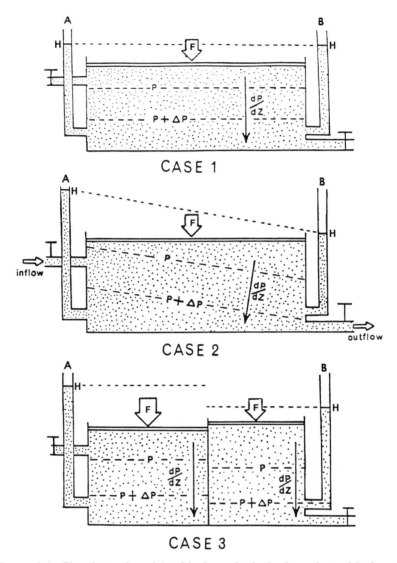

FIGURE 3.4. Closed-vessel models of hydrostatic, hydrodynamic, and hydrostatic but independent environments showing pressure gradient, isopressure plane traces, and hydraulic "head" differences that characterize the different conditions.

vertical manometer-like tubes attached to each end, as shown. There are also an inlet valve on the left end, a lower outlet valve on the right end, and, as in the hydrostatic model (Figure 3.1), a lid that fits tightly against the walls. The reason for the lid is to allow a force to be applied to the fluid in the tank. According to Pascal's principle, the internal pressure in the fluid is duly elevated.

In the first case, the valves are closed, and a load is applied to the lid, which transmits a force to the fluid inside. The extent of overpressuring (above what the pressure would be if the vessel were open) is reflected by H, which is the height of the water column in either of the manometer tubes; H is the same in A and B. The internal planes of constant pressure (isopressure planes), the traces of which are represented by the dotted lines, are horizontal, and they increase in value from top to bottom (P, P + ΔP, etc.). Since the water within the vessel is static, the pressure gradient (dP/dZ) is oriented vertically. Also, the straight dashed line connecting the tops of water columns A and B is horizontal and parallel to the traces of the internal isopressure planes. Since the water column heights reflect the hydraulic head of the water at the tube exit points, it can be inferred that the hydraulic head is constant across the water body. Hydraulic head reflects level of potential energy.

The scenario has been changed in case 2. The inlet and outlet valves are open and water is moving through the vessel from left to right; thus, the fluid environment is dynamic. The traces of the isopressure planes are tilted in the direction of movement of the water, and the pressure gradient, which has to be oriented normal to the isopressure planes, is by consequence nonvertical. At the same time, the height of the water column (and thus the hydraulic head) is greater in manometer tube A at the inflow end than in tube B at the outlet end, and the dashed line connecting the tops of the two water columns is nonhorizontal. Implied by this is that the level of potential energy of the fluid at the inlet exceeds that at the outlet. These effects reflect the nonstatic fluid environment.

Consider case 3, which shows the main vessel divided into two separate chambers, each with its own lid, allowing different loads and forces to be applied to the static fluids within the chambers. The load on the lefthand chamber is greater than that on the righthand chamber, so that the fluid in this chamber is overpressured (Pascal's principle) compared with the righthand one. This is reflected by a taller water column in manometer A than in B. From these observations alone, one might assume a dynamic system such as in case 2, but this would be incorrect because there are actually two static systems under different loads, separated from each other by a barrier.

This is an enigma that confronts the geologist engaged in analyzing subsurface petroleum hydrogeology. When the pressures measured in the same aquifer in two separate wells are accurate but different, the difference can be explained by two scenarios. The first is that the unit is hydrostatic and that there exists a subsurface pressure barrier such as a tight fault, supporting the pressure difference on each side. This is then a condition of "no flow" (J. Bradley, personal communication). The alternative explanation is a single aquifer in full fluid communication but a dynamic one with water moving. This is a condition of "flow," and the problem is that in terms of just these criteria, each case looks the same.

Additional geologic or geophysical information is required to achieve a conclusive interpretation.

A closed tank is somewhat analogous to a subsurface reservoir. In exploratory wells, fluid pressures are measured by gauges at different depths relative to different geologic units. From these measurements, the geologist calculates hydraulic heads and on the basis of these results interprets with reference to these models the subsurface environment (i.e., whether the water is likely to be in motion; if the water is in motion the intensity and direction of flow; and the ultimate effects on the transportation, distribution, and accumulation of any associated hydrocarbons).

Internal Forces—Hydrodynamic Environment

The vector model expressing the major forces operating on unit masses of water, oil, and gas in a dynamic fluid environment is illustrated in Figure 3.5. The only component in this scenario not present in the static case is a

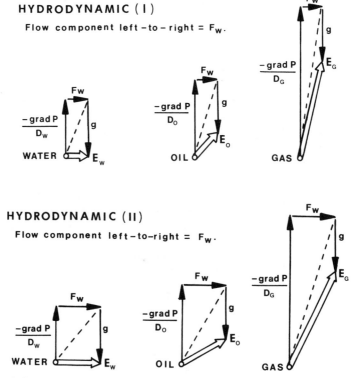

FIGURE 3.5. Vector model showing distribution of principal forces acting on elements of water, oil, and gas in a hydrodynamic fluid environment.

flow vector (F_W), which is oriented horizontally, representing a force directed towards the right. In the first case (Hydrodynamic I), the flow component is relatively small, and the resultant vector (E_W) representing the force acting on a unit mass of water (resolved from buoyancy upward, flow force to the right, and gravity downward) is the white arrow. For the unit mass of oil under the same conditions, the resultant vector is oriented both upward and to the right, because of the positive density-dependent buoyancy component. For the unit mass of gas under these conditions, the resultant vector is very much upward and to the right as well, reflecting the dominent buoyancy force component.

Hydrodynamic II illustrates the same mechanics but in a more intensely hydrodynamic environment. The effect of the increased flow vector (F_W) on the resultant vectors for water, oil, and gas is to pull them more closely into the direction of flow, although the buoyancy effect can never be totally eliminated. Within a subsurface aquifer, of course, the vectors are oriented appropriately in three dimensional space.

Hydraulic Segregation of Hydrocarbons

An important implication emerges from the constructions just discussed; they suggest that in a hydrodynamic regional aquifer it is impossible for the formation water and disbursed elements of oil and gas to follow the same migration paths! And if this is true, one might speculate that over large geologic basin-size areas, like the Powder River, Williston, Big Horn, or Alberta, and others, hydraulic segregation of hydrocarbons at the reservoir level occurs. Free oil must follow one unique hydro-dynamically controlled route with eventual accumulation in geologic features that are capable of holding it. Gas following a hydrodynamically determined different path finds its way into structures where the com-bination of hydrology and geology is favorable for its entrapment, and in this way unique oil and gas reservoir trends develop, not only vertically but geographically.

Hubbert (1953) offers the example shown in Figure 3.6. This construc-tion illustrates a hydrodynamic situation in which the orientation of the forces acting on oil, water, and gas is such that within a monoclinally dip-ping aquifer the gas migrates "updip," while the oil is carried "downdip," even though it is less dense than the downdip-flowing formation water. Notice the traces of planes normal to the individual force vectors. These represent levels of constant potential energy that *decrease in the direction of force application* (since fluids move from regions of higher potential energy to those of lower potential energy). These *isopotential planes* are significant because they can be constructed from pressure measurements to reveal the three-dimensional orientation of the individual force vectors that control the behavior of the oil and gas.

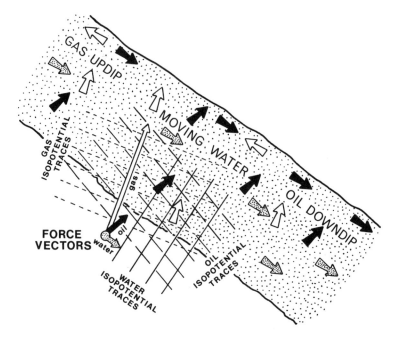

FIGURE 3.6. Migration of oil and gas in different directions as a result of downdip flow in a homoclinal aquifer. Result is hydraulic segregation of oil and gas, a separation process that may create unique oil and gas trends on a regional scale in geologic basins. (Adapted from Hubbert, 1953.)

Figure 3.7 shows the strong influence of a hydrodyamic fluid environment on hydrocarbon entrapment in a simple anticline. The water moves from left to right in response to the flow force produced by the difference in potential energy (higher to the left, lower to the right). The water follows flow paths influenced by forces represented by the stippled arrows. The oil moves in response to those oriented as illustrated by the black arrows and is held in the structure (where the minimum potential energy level point for the oil is located) but *not* at the top of the structure. The gas, moving in response to forces represented by the white arrows, accumulates near the crest of the feature centered on its minimum potential energy point. The oil–water and gas–water contacts are not horizontal but rather tilted as shown, and the attitude of the tilt is *normal* to the orientation of the individual force vectors. Physically, the fluid contacts are planes along which the potential energy levels of the fluids above and below the interface are equal, so that along the boundary the level of potential energy is constant. Thus the fluid contacts are parallel to the isopotential traces and normal to the specific force vectors.

Before tilting can occur, the denser phase (the water) must be moving beneath, and in physical contact with, the lighter, overlying hydrocarbon phase.* The interface between the oil and the gas is essentially *horizontal*, as shown, because to each other these phases are static once they become trapped and held in the structure. All forces are in balance.

If the levels of potential energy associated with moving formation water are mapped, the orientation of related oil and gas–water interfaces can be predicted and the probable migration paths of the hydrocarbon phases can be recognized. These results can predict the regional likelihoods of flushed-out structures, structurally displaced oil and gas pools, hydrocarbon accumulations in nonclosed geologic features, and possible gas occurrences stratigraphically underlying water in deep basins.

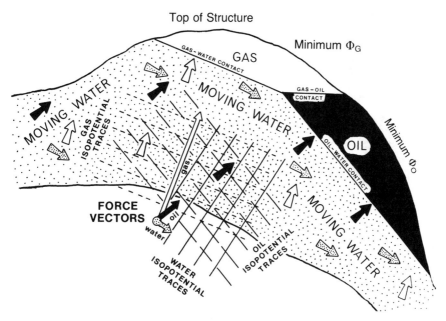

FIGURE 3.7. Effects of a hydrodynamic environment on the entrapment of oil and gas in a simple structure. Note the alignment of the vectors, representing the principal forces acting on the water, oil, and gas and also the traces of the isopotential energy planes and hydrocarbon–water contacts that are normal for each phase.

---

*This is not a frictional pull-down as much as one created by the flow-induced rotation of the invisible but yet influential isopressure and related isopotential-energy planes as described above.

Hydrocarbon Accumulations as a Function of Hydrodynamic Intensity

The comparative effects of hydrogeologic factors on the locations and types of oil and gas accumulations in a subsurface structure are illustrated in Figure 3.8, which shows a cross section through a hypothetical reservoir

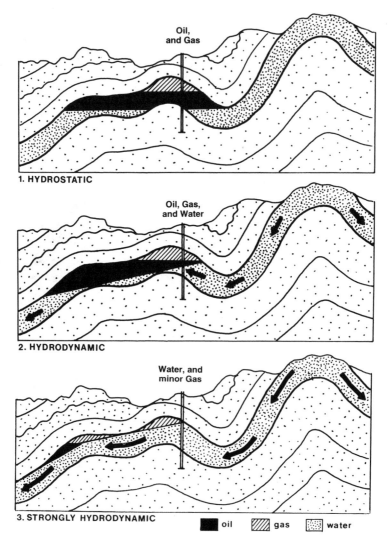

FIGURE 3.8. Cross sections showing the effects on oil and gas accumulations of hydrostatic, moderately hydrodynamic, and strongly hydrodynamic conditions. Fluids recovered by the well shown in the figure reflect the hydrologic environment in the reservoir.

unit. Under static conditions, oil and gas pools are trapped in closed structures (containment from four directions and an overlying seal) such as the one appearing in the figure. The hydrocarbon–water interface is horizontal in the traditional manner, and the well shown in this figure will produce oil and gas. As shown in the second figure, when the fluid environment is dynamic with the water moving in the directions indicated by the black arrows, the effect on the oil accumulation can be quite drastic. The same well would in this case penetrate gas, a reduced column of oil, and water. The thickest oil column would be located in the down-flow direction from the crest of the structure. The distance and oil–water contact tilt would depend on intensity and direction. The gas leg would be unaffected, and the gas–oil contact would be horizontal. Further displacement of the oil would occur if the hydrologic environment became more intense, as illustrated in the third figure. The exploratory well will produce mostly water with a little gas, with the bulk of the gas located in the downflow direction from the well location. The gas–water contact is tilted, as indicated. There would likewise be little indication other than oil staining of the missed offset oil leg, although one would know that hydrodynamic causes can create this effect. It is thus helpful to recognize the kind of fluid environment early in the drilling effort and to be on the alert for hydrodynamic effects so that the nature of the reservoir and the potential production can be correctly understood. For fuller understanding, the characteristics of the hydrodynamic component of the formation can be determined using the methodology described in succeeding chapters.

Effects of Flow Magnitude and Direction on Locations of Hydrocarbon Pools

Since buoyancy is the major force acting on oil and gas within a hydrostatic subsurface unit, the prospective locations of hydrocarbon accumulations will be sites where three-dimensional containment of oil and gas is offered. This is provided by a concave (downward) impermeable reservoir seal on the top and sides and by the buoyancy pressure of the underlying water pushing the hydrocarbons upward from below (see trap model profiles in chapter 4). The potential energy minima under these conditions are located at the highest point in the reservoir.

When conditions are hydrodynamic, factors causing entrapment change markedly in terms of geometry, size, and location of hydrocarbon pools. Increasingly strong hydrodynamic forces caused by mechanisms such as tectonic uplift or downdrop as well as compactional "squeeze", will push hydrocarbons farther and farther from structural accumulation sites until they are totally displaced and the original trapping features are completely filled with flowing formation water. At this point, the traps are "flushed" and attention must be directed at predicting where the oil and gas have gone and where the new accumulation sites might be found.

Figure 3.9 is a map showing the locations of oil pools in a hydro-dynamically benign (hydrostatic) reservoir environment. The pool outlines

FIGURE 3.9. Regional map showing structure on a model reservoir unit. The locations of the oil pools under hydrostatic conditions coincide with the locations of the structural highs, and the pool outlines parallel the structure contours.

coincide closely with the closed structure contours, and an exploration strategy addressed to predicting the locations of and subsequently drilling the "highs" would be both appropriate and probably successful. If factors such as source rock availability, maturation history, and ease of migration were favorable, the oil pools would accumulate at these sites; the potential energy minima toward which all hydrocarbons must move would be located at the tops of the structural traps. As conditions become increasingly hydrodynamic, however, the picture changes.

If we add a regional flow pattern illustrated by the potentiometric map in Figure 3.10, procedures (to be described later) can be applied to modeling the effects of this hydrodynamic component on the oil accumulation site locations. As the figure shows, the regional flow pattern is roughly radial. The map implies that water is entering the reservoir unit along the central anticline via faults or fractures from above or below and spreading out radially within the aquifer toward the west, north, east, and south; the flow paths are indicated by the white arrows. Figure 3.11 is a map showing the regional variation in the potential energy not of the migrating water as in Figure 3.10 but of the oil phase, which is affected by the moving formation water. As the map indicates, under these conditions several nonclosed structures become potential sites for hydrocarbon entrapment because of the downdip flow and the resulting tilted oil–water contacts produced by the dynamic formation water. These accumulations are indicated on the map by the arrows. Notice the *destructive* effect of the moving waters; the three largest original oil pools associated with the major structural highs under static conditions are now flushed of oil and are water filled. In the case of pools A and B (see Figure 3.9) associated with the two anticlines in the north central part of the area (original outlines shown by cross-hatching), the drained hydrocarbons are retrapped downdip in the gray stippled areas. These pool sites are reflected by the closed lows in the contours,* which represent relative levels of potential energy of any oil at any location within the formation water. The available oil migrates toward the lowest points on this map as they represent potential energy minima hydrodynamically displaced from the original stucturally related locations. Some of the other pools in the area have lost portions of their hydrocarbons and have been displaced (although not completely removed from their original locations). Alterations depend on the shape and magnitude of the trapping structure and the direction and flow intensity of the moving water. At the least, the pool geometry is altered slightly; at the most, the entire accumulation is destroyed. The overall effect of these factors on any particular hydrocarbon accumulation can be constructive or destructive, as demonstrated. On the *positive* side,

---

*These are not structural contours but isopotential energy contours that are sensitive to geologic as hydrologic factors. The size of the largest possible oil pool is indicated by the area of the *highest* closing contour!

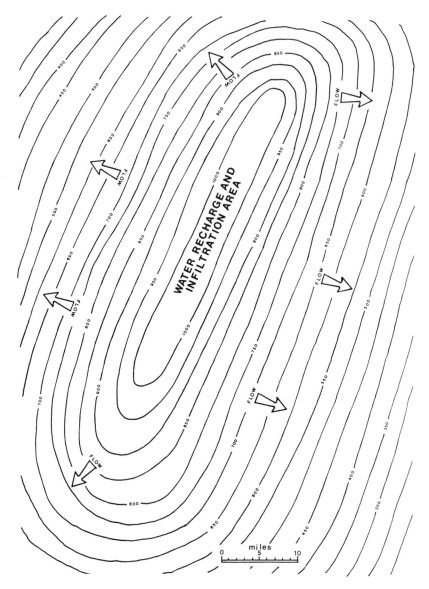

FIGURE 3.10. Potentiometric map reflecting the internal water flow pattern within the model reservoir. Regional flow pattern is essentially radially outward from the center of the area, as indicated by the arrows.

four completely new pools have been created in geologic features that under static conditions would be unable to trap and hold oil.

The hydrodynamic influence has created new pools T, W, X, and Y and has extended old pool I into Z. The purpose of hydrodynamic

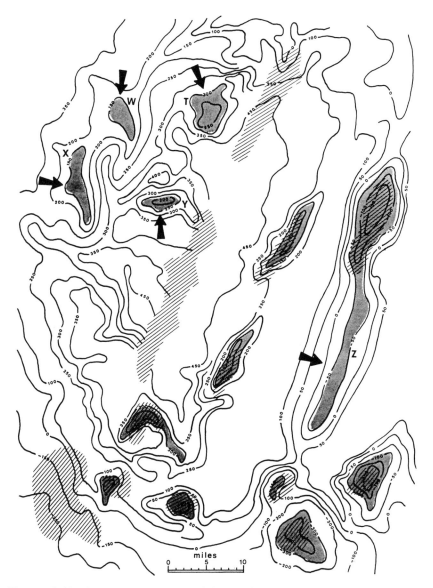

FIGURE 3.11. An entrapment potential map illustrating the displacement and redistribution of oil accumulations, which occur under hydrodynamic reservoir conditions. The contour lines are based on calculated values of the potential energy level of the oil in the moving-water-saturated formation.

exploration analysis is to predict locations where entrapment is *enhanced*, as well as those where entrapment possibilities have been *diminished* by water interacting destructively with the geology. For example, by evaluating the entrapment potential of this area, it might be revealed, prior to costly drilling, that the large domal structure (C) is flushed and that a well on the crest of this feature would be a failure.

Increasing the original hydrodynamic gradient by a factor of four produces almost complete flushing of the closed structures. Figure 3.12

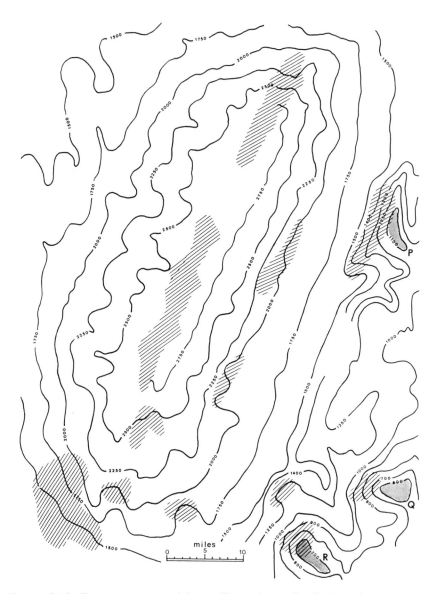

FIGURE 3.12. Entrapment potential map illustrating redistribution of oil accumulations resulting from a more intensely hydrodynamic environment than that illustrated in Figure 3.11. Note that most structures are completely flushed of their hydrocarbons.

illustrates the hydrocarbon entrapment potential for the area at this increased level of flow. As the map indicates, all the original traps shown in the figure as cross-hatched bodies are unable to retain hydrocarbons in this intensified fluid environment. Even the nonclosed structural noses that are stable under less intense conditions, where the buoyancy of the oil and the downdip flow force of the water are in balance, are flushed. The only accumulations remaining in the entire area are the three small hydrodynamically held accumulations in the southeastern corner, indicated by the stippling. R, Q, and P are closed contour lows on the oil entrapment potential surface toward which, according to theory, any hydrocarbons in the area would have to move in response to the potential energy gradients and then become trapped within. Exploratory wells drilled on structural highs in such a hydrodynamic environment would produce water since liquid oil accumulations could exist only on the downdip flanks of a few of the structures as shown.

Further amplification of the regional flow scenario to six times the original represented by the potentiometric map in Figure 3.10 produces the oil entrapment picture in Figure 3.13. In this hydrodynamically intense environment, there are no oil accumulations possible, as indicated by the absence of closed-low-type anomalies on the map. The force of the moving water at this level completely overcomes the buoyancy force of the less-dense oil draining all of the geologic sites. The actual migration paths followed by the hydrocarbons coincide with the channel-like lows on the map, which emanate from the cross-hatchured pattern sites of the original pools.

The theoretical significance of this sequence of cases (the bottom line) to the exploration geologist is that for predicting probable locations of oil and gas accumulations in such an area it is helpful to study the regional hydrogeology as well as structure and stratigraphy. Once the relative intensity of the flow regime, directions of flow, and related factors have been appropriately mapped, the results can be developed into an overall subsurface picture to discern the extent to which hydrocarbons have been dislodged from or even prevented from entering and being retained by local geologic features. This leads to strategic exploration decisions on whether downdip accumulations might be sought on incompletely closed features such as structural noses and whether wells located on terraces, hinges, or along fault planes might yield acceptable amounts of hydrocarbons. On the other hand, if the hydrodynamic entrapment analysis suggests that the regional flow history is likely to have been so intense as to have precluded the successful entrapment of hydrocarbons because of flushing or the diverting of hydrocarbons away from otherwise desirable subsurface features, then exploration effort can be redirected to more hydrogeologically favorable areas. The operational and analytical procedures to be covered in subsequent chapters are the means for assessing these interacting factors, constructing the appropriate maps, and creating the desired leads to future discoveries.

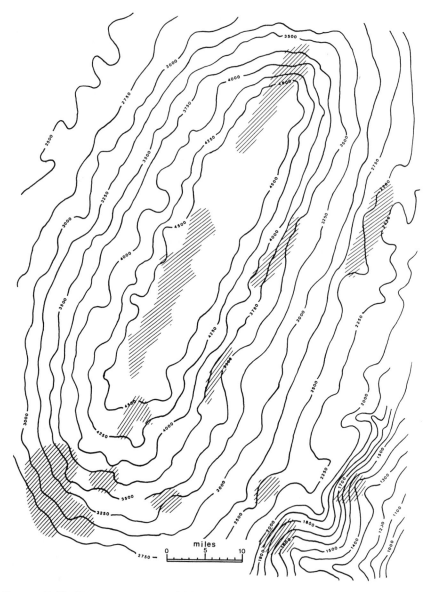

FIGURE 3.13. Entrapment potential map illustrating the advanced destructive hydrodynamic phase. All traps have been flushed in this extreme environment. Water flow paths are normal to the individual contours, and direction is from high potential energy levels to lower ones.

Characteristics of Hydrostatic Environments

The following are the major characteristics of hydrostatic environments:

1. There is no fluid movement (flow).
2. Pressure gradient is vertical toward the center of the Earth increasing at a rate proportional to the density of the fluid concerned.
3. Internal planes of constant pressure (isopressure surfaces) are oriented horizontally.
4. Gravity-density-dependent buoyancy is the only force (vertical) acting on hydrocarbons dispersed within the formation water.
5. There is no significant difference in hydraulic head level from point to point within the formation.
6. There is no variation in internal fluid potential energy level within the body of fluid.
7. Contacts (interfaces) between immiscible fluids of different densities are horizontal (normal to the pressure gradient).
8. Hydrocarbons migrate to the highest points on structures, and accumulations grow from this point (in three dimensions) outward and downward. Existing accumulations shrink toward this point from below and inward from the pool edges.
9. Oil and gas accumulations occur only in geologic features that provide confinement from four sides and the top.
10. Differences in hydraulic head reflect the presence of pressure-resistant subsurface barriers.
11. Outlines of oil pools conform to and parallel local structure contours.
12. Oil and gas follow the same migration paths but move at different rates.
13. The potentiometric surface for the water is horizontal, as are internal planes of constant hydrocarbon potential energy.

Characteristics of Hydrodynamic Environments

The following are the major characteristics of hydrodynamic environments:

1. Fluids are moving within the unit.
2. Pressure gradient is oriented nonvertically because of effects other than just density and gravity.
3. Planes of constant pressure (isopressure planes) are nonhorizontal, tilted downward in the direction of flow.
4. Hydraulic head levels vary from place to place within the fluid body.
5. Level of potential energy varies within the fluid body, diminishing in the direction of fluid movement.
6. Interfaces (contacts) between different fluid phases are nonhorizontal, tilted in the direction of fluid movement or potential energy decrease. Magnitude of dip reflects intensity of flow.

7. Oil and gas accumulations will not always be located close to the highest points on appropriate structures.
8. Hydrocarbon pool outlines will *not* match and parallel the local structure contours.
9. Hydrocarbons can accumulate in geologic traps that are not completely closed or sealed because the force of moving water reinforces weak physical barriers, and holds hydrocarbons in place.
10. Differences in hydraulic head reflect water flow from high to low.
11. Oil and gas migration paths are not similar, because of density differences and water movement.
12. Potentiometric surface is non-horizontal and variable.

4

Potential Energy Variation in Fluids

Potential and Kinetic Energy

When a spring is compressed, work is performed on it to squeeze it together. A portion of this work energy is stored up in the spring as *potential energy*, which is used up (converted to *kinetic energy*) as the spring expands back to its original, presqueezed length. Potential energy is thus energy made by storing up the results of some active work. One does work in winding up tighter and tighter the spring inside a small toy car. When the car is released on the floor, the stored energy converts to kinetic, which works against inner friction of the car's driving mechanism, the friction of the wheels against the floor, the car against the air, and so on, and propels it some distance. The distance depends on how much potential energy was stored up by the spring in the first place—how tightly it was wound or how much work was originally done on it.

As everyone knows, it's a lot of work running up a hill; this is because one is elevating one's level of potential energy (stretching the "spring" of gravity) by moving one's body farther away from the center of the Earth. Potential energy is being stored as you go higher and higher, and this converts back to kinetic energy on the downside of the hill. It's easier to trot or roll down to a lower level with the gravity "spring" pulling you back down. In summary, kinetic energy is energy due to its motion (e.g., the sting of a rubber band), while potential energy is possessed by a body as a result of its position and the work done to get it there (stretching that band!)

Consider a ball placed on a window sill fifty stories high and another on a ledge ten feet above the ground. Relative to ground datum, the mechanical energy possessed by the first ball as a result of the work done to raise it to that great height would very much exceed that of the second; the amount of energy required to lift that ball to the fifty-story height is many times more than that needed to place the second ball on its low-level perch. If each ball drops, the nature of its energy immediately converts from potential to kinetic, and an individual on the ground who had the misfortune of being struck on the head by each could easily discern from the intensity of each blow (i.e., work done on the skull!) which of the two had fallen from the greater height (in other words, which of the two had originally possessed the greater degree of potential energy). In the same way, levels of energy potential at various points within a body of fluid can be mapped, when dealing with subsurface reservoirs. The units are height or elevation in feet or meters relative to a convenient datum, usually sea level. A pressurized liquid has stored energy associated with it, in the same way as a ball raised to a high perch has the potential to do work.

Hydrodynamic theory (developed largely by M.K. Hubbert)* and its applications in exploration geology require being able to recognize and characterize aspects of subsurface fluid behavior in terms of the comparative levels of potential energy reflected by hydraulic head of the fluids concerned. The hydraulic potential can be thought of as the relative level of energy possessed by a unit mass of that fluid at any given point within it. The subsurface reservoir environment is basically created by the forces of gravity and externally applied forces. As discussed earlier, flow or movement occurs as a response to imbalances between regions of high potential energy and those of low potential energy, as illustrated by the simple model in Figure 4.1.

This model consists of a marble rolling on a sloping, tilted surface. The surface variations on the track are analogous to energy levels. When a horizontal force in the form of a push is applied to the marble at the top on the incline (level 11), it will roll down the incline in response to this force combined with the downward pull of gravity, coming to rest in one of the troughs (A at level 9, B at level 6.5, C at level 3.5, or D at level 1.5, depending on the magnitude of the original "push"). Once the marble comes to rest in one of the "lows," its kinetic energy level (motion related) drops to zero, and its potential energy (a function of height) will be a locally minimal value. The marble in this position is locally immobilized, "trapped" in a local low, and it remains there until a new force sufficient to push it up (against gravity!) and over the barrier is applied or the barrier is somehow eliminated. Once the "high" is overcome, the marble continues to move downhill until it comes to rest in another trough because of another obstacle that cannot be surmounted by the kinetic energy of the marble. Of course, as the marble climbs up a high, the kinetic energy of motion that it possesses converts to potential energy, which at the bottom of the low (at which point the kinetic energy is at a maximum) is locally minimal. The traces of the contour lines in the

---

* Hubbert's quest was to formulate a hydrological equivalent to Ohm's Law (E = IR) in electricity, in terms of voltage and current resistance. Fluid analogs are hydraulic head, transmissibility of the rock (water transmission ability) and flow intensity (force).

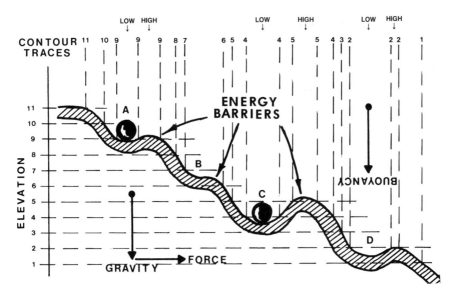

FIGURE 4.1. A mechanical model illustrating local energy barriers and potential energy minima, friction, work, and force in terms of a gravity-driven marble rolling down a sloping irregular track.

figure provide a kind of picture of the surface, indicating the locations of the highs and the lows. The behavior of the marble and the barrier-related points where it might become trapped can be predicted from examining the contours.

If Figure 4.1 is inverted and interpreted as a globule of oil or natural gas (with a density less than water) enclosed within a brine solution, the gravity vector becomes a gravity-density-difference-related buoyancy vector representing force directed upward. This energy potential model is relevant to subsurface hydrocarbon entrapment. At locations A, B, C, and D, the hydraulic potential within the water with respect to the hydrocarbon globules becomes low. The globules therefore come to rest stably in the original lows just as the gravity-driven marble did. Subsurface hydrocarbon traps are in the same way underground regions of locally low energy potential with respect to oil and gas. Hydrocarbons will accumulate within such traps until they gain enough (a sufficient level of) energy (buoyancy force plus some nonvertical "push") to move out from under the restraining barrier.

The application of hydrodynamics to petroleum exploration involves mapping in the subsurface the regional and local variations (even vertical) in fluid potential to predict the behavior of the fluids, just as the simple contour picture of the marble track would allow a viewer to predict the marble's behavior on it.

Potential Energy in a Fluid

It is important that fluid *potential* and fluid *pressure* not be confused because, though related, they *are not the same*. These terms were sometimes used ambiguously and interchangeably, until Hubbert convincingly demonstrated the significance of the difference between them. In formulating the original ideas upon which the concepts of petroleum hydrodynamics are based, Hubbert claims to have been looking for a fluid analog to Ohm's law, which describes the flow of electric current through a wire. Essentially, he sought derivable and mappable hydraulic parameters corresponding to voltage, resistance, and current (E, R, and I in Ohm's law). Since electricity flows from regions of high voltage potential to those of lower voltage potential, Hubbert reasoned that water moving through the pore spaces in a rock must behave similarly and be describable by similar laws and equations. In the same way that a generator does work to "pile" electrons up around the cathode of a battery so that they can eventually flow out to appropriate destinations (such as light bulbs) when the circuit is switched on, potential can be built within a fluid by piling it up above a local datum.

The potential energy of a fluid, which reflects the work it can do, is designated as ϕ and is a function of gravitational force (the gravity "spring"), the height of the point concerned above a datum (the position, or how much the "spring" has been stretched), local pressure produced by external influences, and the density of the fluid concerned. This parameter can be calculated as follows (ignoring capillary force and velocity):

$$\phi = gZ + P/D, \qquad [4.1]$$

where

g = acceleration of gravity
P = pressure
D = density
Z = elevation of the point of interest relative to a reference datum.

This formula and others that follow from it will be used again and again throughout this book; the reader should remember it.

Flow, which is the act of moving along smoothly, occurs in response to point-to-point differences in fluid potential. If the potential energy level at point B is greater than that at point A, the resulting flow direction will be from B toward A for water in an aquifer or a river and in the case of electricity, where electrons migrate in response to a difference in electrical potential (number of electrons or voltage) between two points connected by a wire.

Force Fields, Flow Lines, and Gradients

Figure 4.2 is a contoured map of a surface showing a topography with highs and lows on it. If the map represented a rigid piece of plastic that had been molded into the form illustrated, a marble placed at any location on the surface would roll along a path indicated by the black arrows, from the high down into the lows. And in the same way as the previously described track model, the marble's potential energy would be a function of its elevation or position on the surface (high energy on the topographic "high" and low energy in the "sink"). From most locations, the marble would end up "trapped" in this sink, like a golf ball in a hole. The contours on the surface could be viewed as potential energy contours as well as structural ones, at least for the marbles.

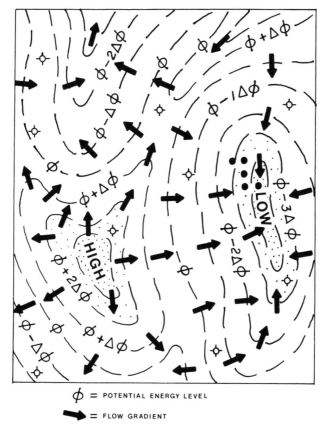

ϕ = POTENTIAL ENERGY LEVEL

➤ = FLOW GRADIENT

FIGURE 4.2. A map on which the contours reflect levels of potential energy. In a fluid environment, the controls from which the "topography" of the surface is inferred are hydraulic head values calculated from water-saturated formation pore pressures.

In the same way, it is apparent that if the level of potential energy within a fluid body in the subsurface can be calculated and mapped on a regional scale, it is possible to define the movements of waters and associated hydrocarbons, in terms of where the fluids are coming from, where they are going, and, more significant, where they are likely to accumulate in temporally stable positions. That is, of course, if the environment is hydraulically dynamic! Recall from the vector models that under hydrodynamic conditions it is physically impossible for the water, oil, and gas to follow exactly the same paths. This makes the problem intriguing because, if this is indeed true, regional hydraulic sorting of hydrocarbons can theoretically occur.

The contours on the map are referred to as *isopotentials*: each represents a specific level of potential energy. The interval between the isopotentials is a constant interval of potential energy designated as $\Delta\phi$. For example, the ϕ values might be 150, 200, 250, 300, etc., so that $\Delta\phi$ is then 50 units. The *flow paths* are indicated by the black arrows, which are oriented normal to the isopotential contours and point from high levels of potential energy to lower levels. They correspond essentially to directions of potential energy decrease expressed as $-\text{grad }\phi$. Thus, by contouring potential energy within an aquifer, flow directions (black arrows) can be inferred. No isopotential contours would show up on a hydrostatic aquifer since it is *static* body of fluid with potential energy at a constant level throughout it, and there is no fluid movement (by definition).

Figure 4.3 illustrates such a static body of fluid within which there exists a fluid pressure gradient increasing from the surface, where the pressure is 1 atm, to the bottom where p is greater. The horizontal lines are the traces of the isopressure planes, and the increment is Δp. The white arrow represents the pressure gradient (grad P) which is constant from top to bottom. A body with a different density (D_H) and volume (v) is submerged in the surrounding fluid. The small p vectors surrounding the immiscible body represent magnitudes of hydrostatic pressure pressing from the surrounding fluid onto the body (greater at the bottom than at the top). The result of these pressures is the upward-oriented buoyancy force (D_Hg), B, which if it exceeds the weight of the body (vD_Hg) will cause it to rise (positive resultant, R). D_H is the density of the hydrocarbon phase, and D_W is that of the water. In a hydrostatic body of fluid at rest, these are the principal forces.

Force Models of Hydrocarbon Traps

A hydrocarbon trap is a three-dimensional space in the subsurface within which hydrocarbons accumulate and are held by whatever forces can accomplish this. Generally, the forces that contain hydrocarbons from above by resisting their natural buoyancy are capillary forces related

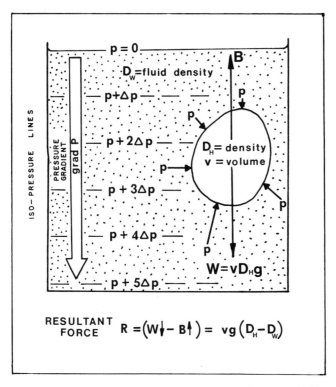

FIGURE 4.3. In a hydrostatic environment, the only force available to move hydrocarbons into geologic traps is buoyancy. The relationship among volume, density, pressure, and buoyancy in a two-immiscible-phase system is illustrated in this figure.

to grain size and impermeability which create an insurmountable entry barrier. Those that press hydrocarbons from below are density-related buoyancy forces. Figure 4.4 represents cross-sectional slices through three different kinds of hydrocarbon reservoir. In the first, the containment forces from above are oriented downward and inward from the sides along a surface that is convex upward. Those from below are arrayed normal to a horizontal plane, and their intersections keep hydrocarbons from escaping from the sides. In the second case, the configurations of the surfaces normal to which the forces are oriented are different. The overlying surface is nose-like and the lower one is dipping so that the normal forces are nonvertical. This tilt creates the intersections that close the interior of the trap and prevent hydrocarbons from escaping. If the buoyancy forces from below were oriented vertically (thus normal to a horizontal plane, as in the first case), there would be no intersection on the left side and no hydrocarbon containment. An active hydrodynamic component is required to create nonvertical buoyancy forces.

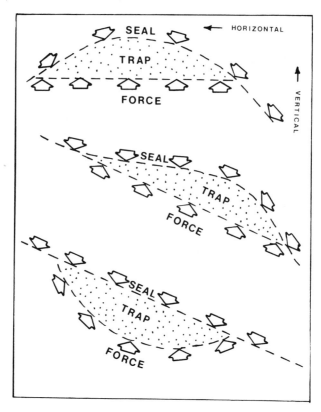

FIGURE 4.4. Orientation of forces that hold oil and gas in various kinds of geologic traps showing configuration of surfaces normal to which the forces acting on the hydrocarbons are oriented.

In the third case, the overlying capillary forces blocking the hydrocarbons from above are arrayed normal to a slanted planar surface, with the trap being completed by underlying buoyancy forces that are arrayed along a concave upward plane that intersects the overlying plane and thus completes the required closure required to hold hydrocarbons.

The first case is the most common; traditional structural and stratigraphic traps with horizontal hydrocarbon–water contacts are of this type, and the capacity depends on the vertical dimension (closure) supplied by the seal. The second case is less common since hydrodynamics are required and the magnitude of the accumulation (hydrocarbon column) depends on both the degree of dip on the oil–water contact and the volume enclosed beneath the overlying impermeable seal. The third type is rare since bending of the hydrocarbon–water contact first down and then up is required to create a space for the oil or gas. This can be accomplished by a combination of hydrodynamics and lithologic barriers.

The contribution to entrapment from the overlying, homoclinally dipping sealing (and ceiling!) lithology is minimal in this case. To create the last two of the three models, flow of the formation water is required; thus, these are essentially "hydrodynamic traps."

Figure 4.5 shows profiles for these three trap models, indicating the contoured levels of potential energy for the hydrocarbons in the water-saturated reservoir for each of the cases discussed earlier. Significant are the gradients, particularly the steep potential energy gradient that is associated with the entry pressure (the pressure that an element of oil or gas has to be able to bring to bear to squeeze into the tighter overlying rock) at the interface between the reservoir and the seal.

Figure 4.6 shows a schematic magnification of this interface (after Hubbert, 1953), along with a graph showing variation in potential energy

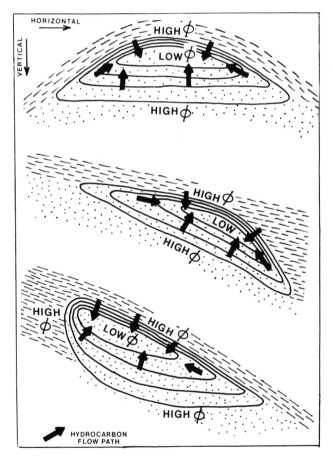

FIGURE 4.5. Levels of potential energy related to the force configurations in Figure 4.4 for three basic hydrocarbon traps.

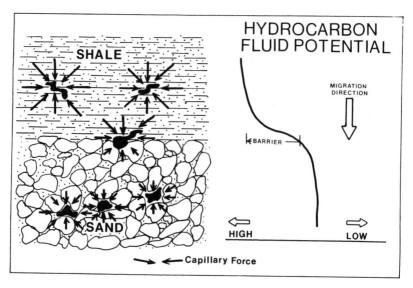

FIGURE 4.6. The shale–sand interface as a potential energy barrier, which is always out of equilibrium with one-way transport (expulsion) from the less porous/permeable source rock to the more porous/permeable reservoir rock.

from inside the impermeable seal to inside the porous and permeable reservoir. The capillary forces acting on globules of oil (assuming a water-wet rock) in the shale would produce compression and elevated internal pressure, as indicated in the diagram. This is analogous to shoving a 2-ft-diameter beach ball into a shoe box (assume an indestructible skin on the ball) and putting the lid on to hold it in. Since much work is done compressing the ball to a smaller size so that it will fit into the box, the level of potential energy for the ball in its confined state is extremely high. The ball would shoot out of the box when the lid was removed, the potential energy converting instantaneously to kinetic. Oil migrates un-directionally from tight rocks into open rocks, where it remains until it migrates farther in response to other forces. The energy environment in the reservoir is lower and less intense, analogous to placing the beach ball in a garbage can. Little work is required, the walls exert minimal force on the ball, and if one raises the lid, the ball probably remains in the can until it is pried out. The level of potential energy for this ball is low, just as is the potential energy environment for a less pres-surized (squeezed) oil globule in a carrier bed or reservoir. As observed by Hubbert (1953), the capillary (displacement)* pressure for oil in shale-size pores is *tens* of atmospheres, while in sand-size pores it is *tenths*. As the curve in the figure reflects, the interface is a steep potential

---

* For oil to physically displace the water.

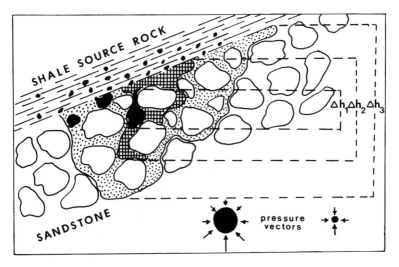

FIGURE 4.7. Schematic migration model suggesting that oil or gas movement within the reservoir or carrier beds occurs only when hydraulic head difference between the top of the hydrocarbon slug and the bottom translates into sufficient buoyancy force to move it through pore throat constrictions.

energy gradient with a drop from high to low between the seal and the reservoir, constituting a one-way barrier. The tendency for any hydrocarbon is to exit the shale unless enough work can be done (force brought to bear) to push it back over the entry barrier by increasing its potential energy sufficiently to overcome the entry pressure (resistance) of the sealing lithology.

Migration occurs within a carrier bed when a body of oil or gas becomes large enough so that the difference in hydraulic head and pressure in the surrounding water between the top and bottom is sufficient to overcome the resistance of the grain throats. The net force imbalance propels the hydrocarbons upward, as illustrated in Figure 4.7.

Pressure, Potential, and Hydraulic Head

Figure 4.8 illustrates an apparatus that has been used to investigate and illustrate the physical and functional aspects of water flow through porous media that as a function of permeability, fluid viscosity, sedimentary grain size, and other factors are interrelated through Darcy's law. As illustrated, the equipment includes a movable (up and down) reservoir to vary the flow force (hydraulic head), which feeds a flow tube of length (L) and cross-sectional area (A). This tube is packed with sediment of particular type or grain size, and there are two vertical manometer tubes rising out of the flow tube, as shown. The water that percolates through the sediment-packed tube exits into a graduated cylinder so that change of

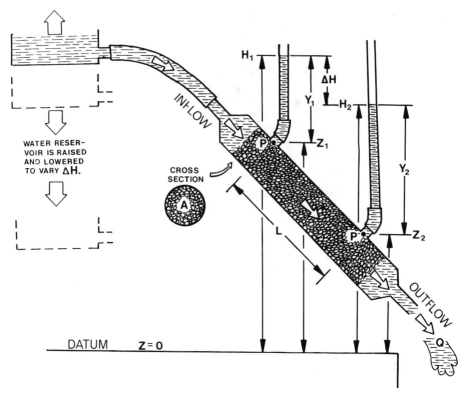

FIGURE 4.8. Diagram of the flow tube experiment that imitates Darcy's original nineteenth-century work on permeability. (Designed by M.K. Hubbert.)

volume with respect to time can be easily monitored as discharge flow rate.

With water flowing through the system, the pressure P_1 at Z_1 at depth Y_1 below the top of the water column H_1 is calculated as

$$P_1 = P_0 + D_wg(Y_1), \qquad [4.2]$$

let $P_0 = 0$, so that

$$Y_1 = \frac{P_1}{D_wg}, \qquad [4.3]$$

and

$$\text{grad } P = D_wg, \qquad [4.4]$$

so that

$$Y_1 = \frac{P_1}{\text{grad } P} \qquad [4.5]$$

	C	D	E	F	G	H
1	PERIMENT – RAW DATA					
2						
3	VOLUME	TIME	FLOW	h1	h2	h1 – h2
4						
5						
6	3	251	1.048437	31.9	29.9	2
7	5.6	234.9	2.109728	37.6	30.3	7.3
8	10	225.8	3.884822	48	33.1	14.9
9	10	177.3	4.947506	52.6	33.8	18.8
10	10	151.4	5.793877	57.7	35.8	21.9
11	10	147.4	5.951105	62.2	39.4	22.8
12	25	312	7.028789	66.7	40.2	26.5
13	25	233.3	9.399839	77	41.8	35.2
14	25	196	11.18868	85.1	43	42.1
15	25	163.4	13.42094	95	44.8	50.2
16	25	142.9	15.34627	103.5	46.2	57.3
17	25	108.1	20.28660	97	22	75
18						
19						
20						

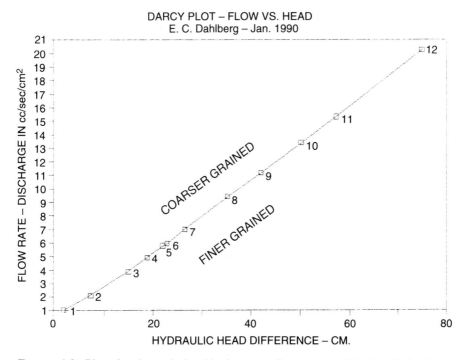

DARCY PLOT – FLOW VS. HEAD
E. C. Dahlberg – Jan. 1990

FIGURE 4.9. Plot showing relationship between flow rate and hydraulic head difference based on twelve "runs" on the apparatus illustrated in Figure 4.8. For each run, the height of the reservoir is set at a different level to vary the hydraulic head.

and since (from figure)

$$H_1 = Z_1 + Y_1$$ [4.6]

then

$$H_1 = Z_1 + \frac{P_1}{grad\ P}$$ [4.7]

which expresses hydraulic head as a function of pressure and pressure gradient. The same logic yields hydraulic head H_2 as a function of P_2.

Since the flow is obviously from the top of the tube where H_1 exceeds H_2, and P_2 is greater than P_1, it is apparent that the direction of water movement is from high hydraulic head to lower rather than from highest pressure to lower. The difference in hydraulic head from tube 1 to tube 2 is ΔH, which is equal to $H_1 - H_2$, either calculated or measured.

When the reservoir is moved to other levels, measurements of the discharge rate are recorded for each, the hydraulic head elevations relative to the illustrated datum are measured, and the difference between the two is plotted against discharge, the result is a graph of the type shown in Figure 4.9. The relationship between flow and hydraulic head drop (which represents loss of potential energy of the water along the flow path due to friction and other factors) is linear.

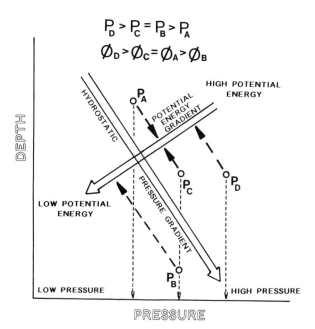

FIGURE 4.10. Fluid pressure–depth differences and levels of potential energy that they reflect. Pressure varies in the horizontal direction, and potential energy varies along a gradient that is normal to the hydrostatic pressure–depth gradient.

For each sediment of a particular grain size, the points plot on a single straight line; the hydraulic head difference is essentially proportional to the rate of flow through the grains for each. The slopes differ, however: the more steeply sloping lines reflect the coarser sediment types, and the less steep slopes reflect the tighter, more resistant, shalier lithologies.

When the pressure at a certain depth in a body of fluid is other than hydrostatic, internal differences in potential energy exist so that the fluid flows and is hydrodynamic. On a simple pressure–depth graph such as that depicted in Figure 4.10, the relationship between pressure and potential energy is illustrated. In effect, a potential energy gradient normal to the static density gradient can be seen. The individual pressure values that project onto that gradient up and to the right of the static gradient reflect fluids from a higher potential energy environment than those that project farther down and to the left. In the figure, P_D is the highest pressure, and the potential energy level of the fluid at that point is also the highest of the four points. P_C and P_B are equal pressures, although the potential energy level of the former significantly exceeds that of the latter, and if the two locations were physically interconnected, fluids would move from the vicinity of C toward the location B. The hydraulic head elevation calculated from P_C is higher than that calculated from P_B. P_A is the lowest of all four, but the potential energy for A is the same as for C and higher than at B. If the two were in fluid communication, the flow would be from A toward B, with no flow between A and C since they project to the same position on the flow gradient (no head difference). It can be demonstrated by calculations that potential energy is constant at all locations within a static body of fluid.

5

Rock, Pore Space, and Fluid Systems

Rocks and Water

The principles and concepts concerning fluid behavior discussed in earlier chapters have centered on the hydrologic component of the subsurface environment in terms of single bodies of homogeneous fluid. It is not really this way in a rock, however, since any single fluid mass is essentially interspersed within a network of intergranular pore spaces, throats, fracture openings, etc. As long as the fluids are in complete communication, the principles are valid, but to understand pressure and flow relationships the effects of the mineral grain matrix must be considered as well.

Sedimentary rocks are pretty much saturated with formation water from about the water table down to depths that have been penetrated by the deepest of oil wells (greater than five miles). The pressure within the pore spaces of the rock is at least hydrostatic (i.e., a function of the weight of the fluids above the point of measurement). This is true as long as the mineral grains can support the weight of the grains above them without collapsing.

Some basic aspects of water-saturated grain systems are illustrated in Figure 5.1, which shows two vessels, each sitting on the pan of a scale. The first vessel contains only water, and the internal fluid pressure illustrated by the white arrows is greater at point B than at point A, a function of the hydrostatic gradient, grad P. The overall force that this system transmits to an underlying one is equal only to the weight of the water (ignoring the weight of the vessel itself). The black arrows represent this force exerted by the bottom of the container against the pan of the scale.

In the second container, rock grains have been added to the system. They create a matrix network that is self-supporting, so there is no difference in the internal fluid pressure, which is now the pore pressure measured within the spaces between the grains. Those at depths C and D are the same as at depths A and B in the first container, a function of the

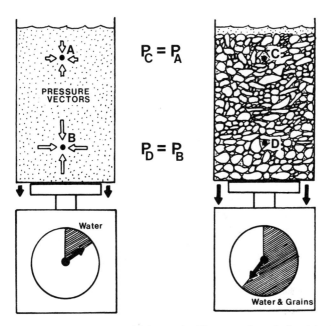

FIGURE 5.1. Two containers on weighing scales illustrate the relationships between internal pressure in an aquifer, external load, and the translation of force as weight to underlying materials.

weight of the water only. However, because of the added mass of the grains, the overall force that is pressing down on the scale exceeds that outlined in the first case. If this weight is transmitted to an underlying body of fluid, the result is an increase in the fluid pressure to a level that exceeds considerably that in the upper vessel. This grains-plus-fluids weight is an external force that elevates the internal pressure according to Pascal's principle, creating a discontinuity in the overall hydrostatic gradient. These principles explain the mechanics of compaction gradients, abnormally pressured formations, and geopressure phenomena encountered in drilling.

Consider the scenario in Figure 5.2, which shows a vessel filled with water and mineral grains with an impermeable seal in the lower part of it. Above the seal, the grains support themselves as a network of struts, so that any pressure measured within the water in the pores is hydrostatic and the gradient along which P_A, P_B, P_C, P_D, and P_E lie is related only to the density of the fluid. Below the seal, the grains are less competent and cannot support the weight of the overlying rock and fluid, so some of the stress is transferred to the *confined* intergranular fluids. As described by Pascal's principle, the internal pressure is elevated above the normal gradient proportionally at all points within the subseal sediments, as

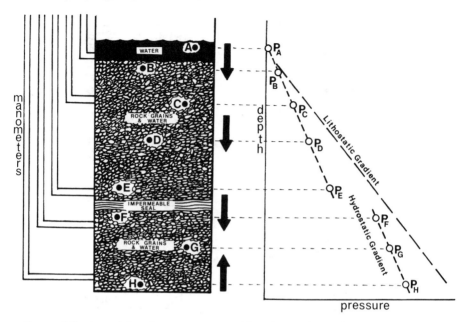

FIGURE 5.2. A model rock–water system with an internal seal that is not supported by the rock grains matrix framework in the lower compartment, within which the fluids (for all practical purposes) are confined.

indicated by the P_F, P_G, and P_H gradient in the figure. The position of the seal is indicated by the break in the gradient on the graph.

Compaction, Porosity, and Pore Pressure

If a column of typical sediment made up of mineral grains is considered in detail, it can be noted at the top of the column that the grains are of predeposition size and shape, and the water-filled pore spaces are a maximum size, as illustrated in Figure 5.3. The pressure within the fluid, is hydrostatic since the sediment is grain supported. Farther down in the column, in response to the weight of the overlying grains, the grains begin to collapse and encroach on the pore spaces occupied by the fluids. If the fluids can escape in response to the force exerted by the grains, the internal pressure is maintained close to hydrostatic, and the sediment is described as partially compacted. Deeper still in the section, this natural grain collapse continues and the pore spaces close up even more tightly. If the fluids can continue to adjust, the internal pressure buildup is held to a minimum and the pore fluid pressure remains near hydrostatic. Throughout this sequence, there is a fairly systematic decrease in porosity. This is described as a "normally compacted" sediment.

If obstacles to migration, such as bounding impermeable barriers or sealing lithologies, prevent fluids from escaping into less stressed regions in response to grain squeezing, the internal fluid pressure increases to a level in excess of density-dependent hydrostatic for the depth concerned.

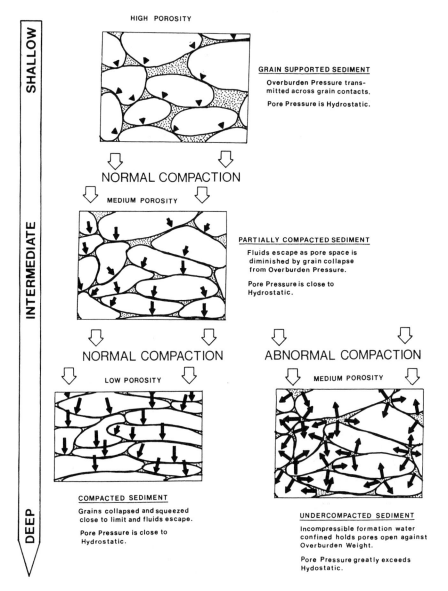

FIGURE 5.3. Changes in pore size and grain shape that occur as an unconsolidated sediment is buried and compacted by burial and the anomalous undercompaction that occurs when pore fluid escape is hampered.

At the same time as the pore pressure rises, the pore space decrease is resisted proportionally, so that the porosity remains anomalously high for its depth and the pore fluids are highly overpressured. This condition is referred to as "abnormally compacted."

A compaction curve is a graph showing the rate of change of porosity with depth. The general form resembles a decreasing exponential function. Figure 5.4 illustrates a collection of published compaction curves, showing the variation in shaley sediment pore volume attributable to differing lithology and other factors. Porosities at depths of less than 1,000 ft. vary from 25% to 60%. At burial depth around 3,000 ft., the porosity range has decreased to less than 10% up to around 30%. At 5,000 ft., the porosity range has decreased and narrowed even further, to 5–25%, and from 5,000 ft on down the changes appear insignificant as the curves level off. It appears that at some limiting depth, the physical properties of the grains and the pore fluids are such that the overburden stress collapses them no further. The depth at which this occurs is a function of the competency of the grains, the composition of the fluid, and the ease of escape of the fluids. The actual lithologies of the sediments represented by these curves are not known, so the broad range in depth–

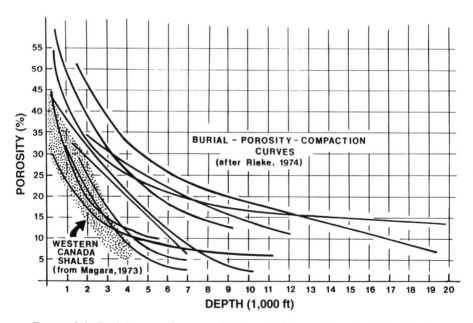

FIGURE 5.4. Burial curves for several different shales and sandy shales showing the rate at which porosity has diminished as depth of burial and thickness of overburden has increased.

porosity curves cannot be uniquely attributable to lithology on the basis of this data alone.

Figure 5.5 shows a hypothetical scenario similar to that in Figure 5.2 except with respect to support of the seal by the underlying grains or by its own mechanical rigidity. The seal is largely supported and therefore transmits little of the weight of the overburden and fluids to the fluids below the seal. In the event of no force acting upon the underlying fluids, the pressure at point D immediately beneath the seal is even less than hydrostatic. This is almost like the top of a completely different column of fluid, the pressure increases down through it at a hydrostatic rate. Pressures P_D, P_E, and P_F lie on a gradient that is static (density dependent) but displaced to the low pressure side of the "normal static gradient" established from the pressures in the chamber above the seal. A "normal gradient" in the subsurface is a reference construction only.

What happens under these conditions if the seal is *not* impermeable? Experimentation is needed to determine the extent to which pore pressures beneath the seal are increased by "squeeze" if they can escape up through the seal into the overlying sediments. It seems that this type of relief would cause the overpressured cell to eventually dissipate. The environment would most certainly be a dynamic one as long as the pressure excess were maintained, and there would certainly be a strong component of upward-directed flow under these conditions. It seems, as well, that when grains fail at different depths and locations in a sedimentary section,

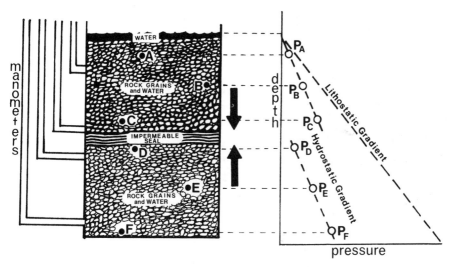

FIGURE 5.5. A model rock–water system with an internal, completely impermeable, seal. the rock framework in the compartment underneath the seal supports the weight of the rocks *and* the water overlying the seal.

the points of maximum collapse essentially function as centers of increased formation pressure and hydraulic head from which fluid flow emanates as long as the stress continues.

Figure 5.6 is a plot of interval transit time in logarithm of microseconds per foot against depth showing different gradient components. (The graph was produced using the Lotus program PGRAPH.) The data are from sonic log picks made at 50-ft intervals in a well from the Atlantic offshore of Newfoundland, and the points illustrate the type of prominent break that is commonly observed near the top of a geopressured zone. The break itself in the compaction trend (gradual decrease in pore volume due to overburden weight) occurs around the 7,000-ft mark, the trend being fairly consistent above this depth, with small deviations reflecting strati-graphic differences in rock type. At 7,000 ft, the general trend abruptly changes, the travel times becoming longer because of lowered acoustical velocity of the sediments presumably due to anomalously high porosity. One would attribute this to an overpressuring in the pore fluids and resistance to the normal compaction with the fluids holding the spaces open. The sonic curve changes can occur as a result of other factors that alter the acoustical character of the rock as well as geopressuring, but such breaks very often characterize the tops of overpressured zones that have been verified with pressure measurements as well.

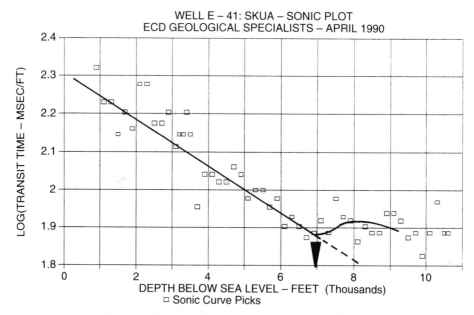

FIGURE 5.6. "Sonitran" graph showing transit time variation over the vertical extent of a well and normal compaction trend with a sudden break (increase in travel time due to anomalously high porosity below 7,000 ft).

Hydrodynamic Rock – Water Systems

Pressure–depth gradients such as those described earlier reflecting seals and discontinuities can disclose that a unit is dynamic rather than static, as demonstrated in the following two models. In Figure 5.7 we have a column of sediment within which the pressure is measured at depths A through H. When plotted on the graph, the values lie along a straight line, implying a single gradient, which is in fact correct. However, if the hydraulic heads are calculated from each pressure, with the head for P_H being the highest and that for P_A the lowest, there is a fluid potential difference from the bottom of the column to the top. From this, one would infer updip flow, with water moving *upward* through the interval. The gradient itself follows a slope that is different from the density-dependent hydrostatic one.

Downdip flow through the sedimentary column is recognized similarly on the basis of a nonhydrostatic gradient trend in the plotted pressures, as demonstrated in Figure 5.8. In the same manner as just described, the hydraulic head* is greater at the top of the rock column than at the bottom, and since flow is fluid migration from higher to lower levels of hydraulic head (in response to a hydraulic gradient) the water is moving downward within this unit.

Water Expulsion and Escape

If an external force is applied to a water-saturated rock grain system, the ultimate pressure buildup within the pore spaces and the duration over which the buildup persists depend upon the extent to which the force-induced pressure is relieved by fluid escape or leakage (either vertical or lateral). Another factor is the point at which grains and the resistance of the porefluids to the applied stress reaches such a level that the sediment cannot be compacted any more. To illustrate this scenario, Figure 5.9 shows a confined system of grains and formation fluid being compressed. In case A, the system is tightly sealed, the water cannot escape, and the pore pressure builds up *in excess of hydrostatic*. The limit is close to the point at which the grains and pore fluids take up the load. Then, if the fluids eventually migrate out of the system, the pressure drops back over time toward the original hydrostatic level. In case B, the pore fluids are not completely confined, and they leak out of the system at a moderate

---

*The further that a pressure value plots from the hydrostatic gradient the greater the hydraulic head for that point.

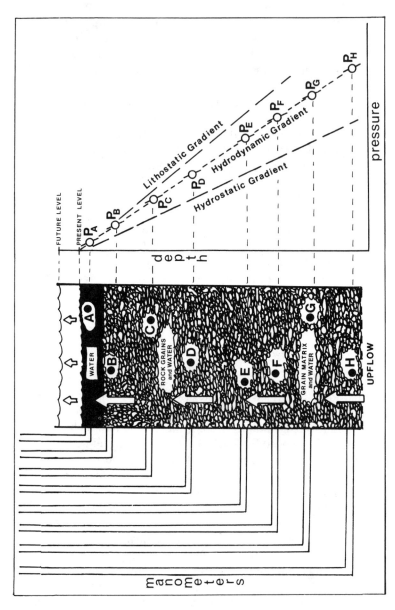

FIGURE 5.7. Pressure–depth relationships and gradients for a mixture of grains through which water is flowing from below as in updip flow.

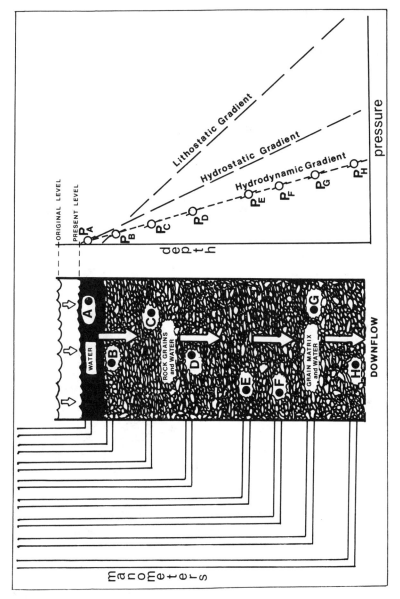

FIGURE 5.8. Pressure–depth relationships and gradients for a mixture of grains through which water is flowing from above as in downdip flow.

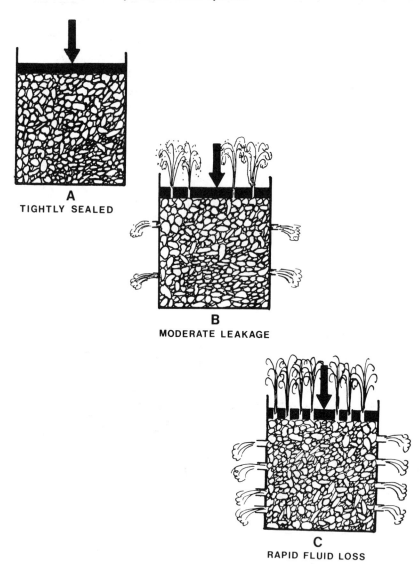

FIGURE 5.9. Rock–water systems demonstrating the effects of squeeze-produced interstitial water expulsion on pore pressure buildup and maintenance as a function of time.

rate as the load force is applied. The buildup is not as high, the grain resistance limit to further compaction is finally reached, and the pore pressure eventually drops back toward hydrostatic. In case C, the system is almost completely open (for fluids), the stress-dependent pressure buildup is smaller and even more fleeting than in case B, and the dis-

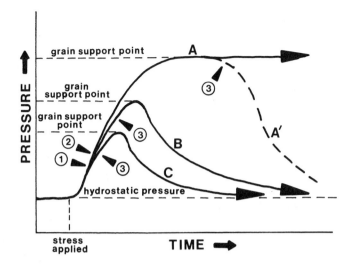

FIGURE 5.10. Graph showing relationships among compaction, pressure, and time for the three cases in Figure 5.9. 1, Force transmitted to fluid; 2, Grains collapsing; 3, Grains supporting the seal.

sipation back to hydrostatic is considerably more rapid. The overpressure episode in this case is extremely limited both in time and magnitude.

Figure 5.10 is theoretical—based on conjecture unsupported by experimentation. The relationships seem intuitively appropriate but are at present unverified. The curves illustrate the time-dependent pressure buildups and decreases that characterize the three cases just described. The numbers represent the stages in the process. The curve representing case A shows rapid buildup of pressure with some grain failure prior to reaching the grain collapse limit and maximum overpressure. Also shown (dashed line) is a gradual dissipation of the pressure if the pore fluid is able to migrate away over time so that the pressure drops back toward the original hydrostatic level. Curve B reflects a pressure buildup to a lower level than A because of a moderate degree of fluid escape in response to the applied force. In case C, the pressure buildup is minimal because the system is almost completely open and the fluids are squeezed out with relative ease and little elevation of internal pressure. The result is attributable to the comparative rates of force application and expulsion of the pore fluids. If one were to measure the pore pressure in the grain–water systems of Figure 5.9 at a particular point in time following the applications of the force to them, as shown in Figure 5.10, P_A would be the highest, followed by P_B and then P_C, the lowest of the three. These pressure differences would be the result only of fluid expulsion rate, since ideally all other factors have been identical. (These ideas will be important in the interpretation of pressure–elevation plots.)

6

Nonnormal Formation Pressures

Abnormal Pore Pressures

An accurately measured formation pressure value that differs significantly from the pertinent hydrostatic pressure for a fluid column from the surface down to the depth of the measurement is termed *abnormal*, even though for the conditions at hand it may be completely normal. The standard is assumed to be the local hydrostatic gradient from the surface, and this gradient is termed *normal*. The relationships for different pressures measured beneath the surface are illustrated in Figure 6.1, which shows the assumed normal gradient standard and some pressure measurements from various units in the section, as well as their corresponding hydraulic head elevations. On the graph, points A and I lie on the normal gradient; on this basis the formations that they represent are considered normally pressured. Points C, D, E, and F lie to the high side of the normal gradient and are interpreted as representing units that are "over-pressured" for the locality. On the other hand, pressures B, H, and G fall to the lower left of the standard normal gradient and are low for their depth, so the units concerned would be classed as "underpressured." These terms are relative and only locally meaningful.

Hydraulic Head and Potential Energy

The relationship between hydraulic head and potential energy was discussed in chapter 4 as part of the Darcy flow tube experiment. Figure 6.2 illustrates that when a gradient with density-dependent slope is projected from a pressure point on the graph, back up through shallower depths to the level that corresponds to a pressure of 1 atm, the elevation (which may or may not correspond to the top of the local water table) is termed the *hydraulic head elevation* and reflects the level of potential energy possessed by the water. A regional map made from a collection of these values is a *potentiometric map*; it is constructed from individual formation pressure measurements.

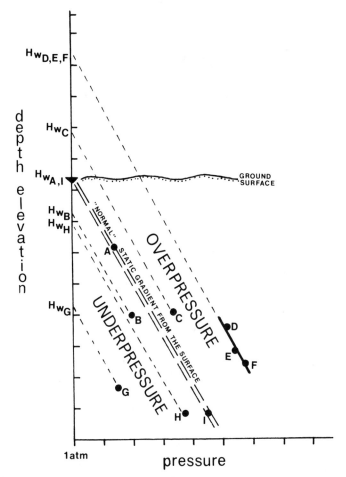

FIGURE 6.1. Pressure–depth gradient diagram illustrating locations of positions of plotted formation pressure measurements and corresponding hydraulic head values for abnormally low, high, and normally (hydrostatic) pressured systems.

Causes of Abnormal Formation Pressure

One cause of nonnormal fluid pressure, illustrated in Figure 6.2, is a formation in communication with the surface at an elevation significantly higher than the ground level at the location concerned (thus it is artesian). In case 1, the outcrop of the formation to the west of the well site has an elevation similar to KB (rig floor elevation) at the well. Assuming a static formation, the measured formation pressure (P_F) would be close to the calculated normal pressure (P_N), the hydraulic head elevation would coincide with ground level, and the formation would be considered nor-

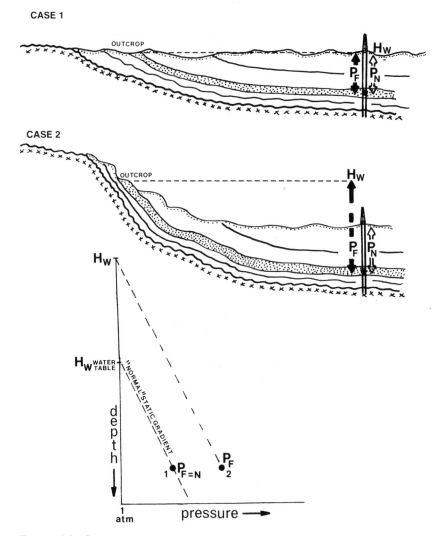

FIGURE 6.2. One source of overpressuring is communication of the reservoir with the surface at an essentially higher elevation than ground level at the well site. Pressure and hydraulic head for the normal and abnormal cases are shown on the P–D graph.

mally pressured for its depth. In case 2, however, the formation pressure (P_F) would greatly exceed the calculated normal pressure (P_N) because of its location deeper in the water column, which is increased by the outcrop of the unit at an elevation higher than the ground level at the well site. The hydraulic head calculated from the formation pressure would be close to the elevation of the outcrop; the pressure points would plot a good distance to the high side of the normal hydrostatic gradient projected as a function of water density downward from the ground level surface at the well. This formation would be highly overpressured, and failure to predict it could produce a damaging and wasteful blowout during drilling. If the water were moving, the formation would be not only overpressured but hydrodynamic as well though the head elevation would be less than outcrop.

Formation pressures that reflect the geology and hydrology of an ancient era can also be considered abnormal because the criteria for a normal unit today do not appropriately reflect the past, when ground levels, climates, ice sheets, and so on, were different and compaction, erosion, faulting, and burial were occurring. Thus, abnormal pressures measured today may in many instances be fossil effects of load, fluid, and temperature conditions in the past. Figures 6.3 and 6.4 illustrate a series of scenarios that demonstrate "inherited" formation pressures in cross section and pressure--depth (P–D) plot form. In (a) we see a unit as it existed at some time in the past under static conditions. The formation pressure then was normal, due to the height of the water

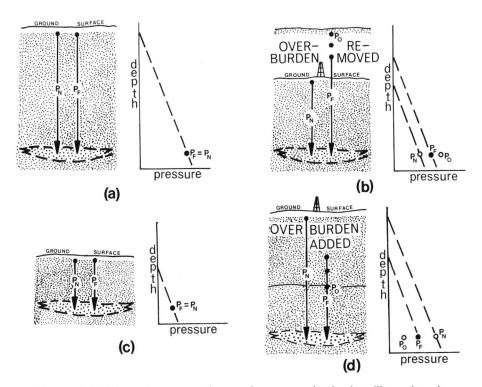

FIGURE 6.3. Schematic cross sections and pressure–depth plots illustrating abnormal pressures that are inherited from earlier conditions. In (a) and (b) are cases of removed overburden, and (c) and (d) show results of added burden.

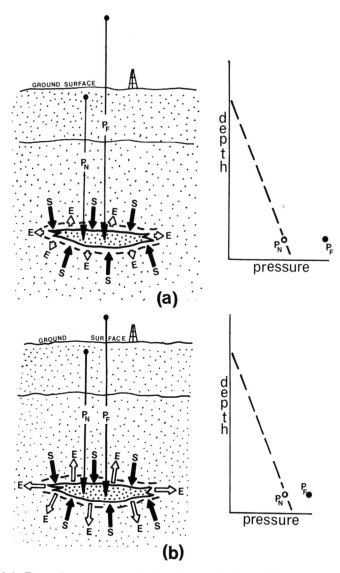

FIGURE 6.4. Formation pressure vairation that results from differing rates of fluid expulsion when squeeze is applied to reservoir fluids. (a) Restricted fluid escape; (b) moderately restricted fluid escape; (c) moderately easy fluid escape.

column (about ground level), and P_N equaled P_N (normal hydrostatic). In (b) we see about the same case following removal of overburden, so that the burial depth has been reduced by erosion. Not having adjusted to these new conditions, perhaps because of its depth, the formation is presently hydraulically out of equilibrium; the original pore pressure (P_O)

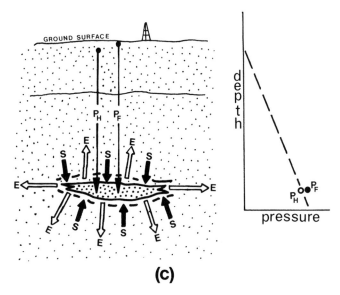

FIGURE 6.4. *Continued*

decreases over time with the diminishing water column and will eventually reach P_N, the static pressure for today's water column. P_F exceeds the criterion normal pressure and the reservoir (as illustrated in the P–D plot) is overpressured for its current depth. Not much is known about the interactions of these influencing factors which have occurred during the huge periods of time over which geologic processes operate.

The opposite can occur, too, commencing with the original condition (c) in which the fluid pressure in the reservoir is due to the height of the local water column at some time in the past and thus hydrostatic and normal ($P_F = P_N$). Over a period of time, as overburden is added (d), burial depth is increased so that at drilling time the formation pressure, P_F, is below the normal hydrostatic for its depth but still greater (as illustrated on the P–D plot) than the original pore pressure (P_O). Note that P_F is in the process of adjusting to the present-day fluid column and might eventually gain the configuration of (a). These conditions relate to the relative speed of geological compared to hydrological processes.

In each of the cases just discussed, formation pressure variations have been created only by changes in the fluid column heights with no effect of overburden rock load, tectonic forces, or other heroic factors. The effects of grain collapse and formation compaction, as discussed earlier, squeeze the fluids in the spaces between the grains to produce pore pressure changes that are independent of hydrostatic water column changes. As illustrated in Figure 6.4, case (a), the load of the overburden above the unit is transmitted to the fluids within the reservoir by collapse of the grains, which cannot support the overburden weight. The magnitude of the stress (or internal squeeze) is represented by the black "S" vectors, and the rate of escape (or expulsion) of the pore fluid in response to the squeeze is shown by the white "E" vectors. In this case, "squeeze" exceeds "escape" so that internal pressure is elevated accordingly, the hydraulic head elevation is well above ground level, and the unit is "overpressured" for its location. P_F exceeds P_N on the P–D graph. Case (b) illustrates a similarly overpressured reservoir, but because the expulsion rate is greater than in case (a), the reservoir fluids are better able to adjust (larger expulsion vectors), so the resulting degree of internal pore pressure buildup is less. The unit is overpressured since the hydraulic head elevation is higher than the ground level elevation. In case (c), the reservoir is close to normal pressure for its depth since P_F and P_N are almost the same on the P–D graph. The reservoir fluid escape in response to overburden stress just about keeps up with the pressure increase, so that buildup and maintenance of the overpressure is minimal. In these cases, formation pressure differences are conceived as being almost completely due to fluid expulsion rate.

Additional mechanisms that produce an excess of confined fluid for the pore space available are illustrated in the next three figures. Figure 6.5 shows the occupation of pore space by mineral overgrowths precipitated onto the surfaces of the grains. If growth proceeds and the resident pore fluids cannot escape as the pore space diminishes, then overpressuring above hydrostatic occurs. The degree of difference in pore pressure and hydrostatic for the depth concerned has not been documented with real data. Diagenetic processes seem to be able to create seals for fluids, however, and mineralization does diminish porosity and permeability.

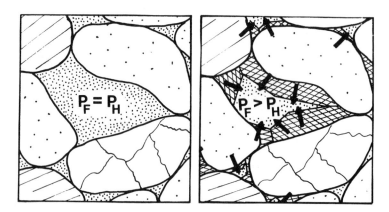

FIGURE 6.5. Reduction of pore space by formation of crystalline overgrowths on the grain surfaces and pore walls elevates the fluid pressure above normal hydrostatic.

Precipitation and growth requires continual replenishment of the mineralizing solution. This could become hampered by the resulting crystal growth, so if this mechanism leads to significant overpressuring the details of the process are quite obscure. Bradley (1975) proposes that excessive heat (which drives precipitated solid phases back into solution and should thus dissolve overgrowths) and lack of significantly supersaturated pore brines in sediments (which enhances precipitation of solids out of solution) may well preclude the possibility of chemically created overpressures or minimize their extent.

However, heating of the overall rock itself should (as illustrated in Figure 6.6) create overpressuring as a result of different rates of expansion within the pore fluids and the mineral grains. Bradley (1975) notes that the volume of a brine increases to almost one hundred times the volume of the pore space that contains it as the temperature of the system rises. Unless the pore fluids can expand into additional spaces, the internal pressure is likewise bound to increase significantly. Bradley furthermore observes that within a confined freshwater system, the pressure can change 125 psi for every degree Fahrenheit change in temperature, and that under normal geologic conditions with sufficient time to establish equilibrium between surface and buried reservoir, a change in ambient surface temperature of 8°F can produce a 1,000 psi deviation from normal hydrostatic pressure within a sufficiently sealed unit, if there occurs no fluid pressure relief by escape.

It has also been proposed that abnormal pressuring is created in an effectively sealed reservoir by conversion of organic matter to gas, oil, carbon dioxide and other volatiles as demonstrated in Figure 6.7 *if* there is a corresponding increase above the original in volume; in other words,

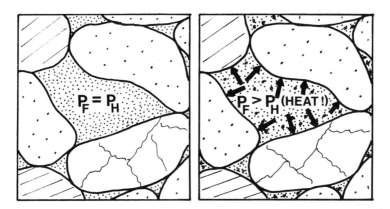

FIGURE 6.6. Heating of the rock increases the pore fluid pressure since fluid volume increase is many times greater than grain volume (and thus pore volume) increase; note the differential thermal expansion of solid and liquid.

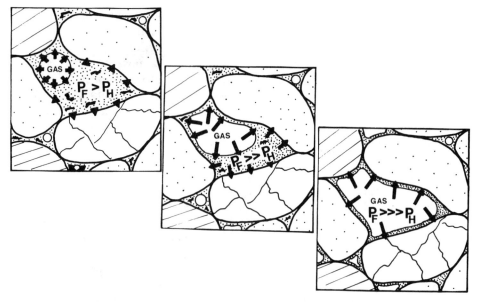

FIGURE 6.7. Pore fluid pressure increases above normal hydrostatic by infusion of gas from nearby coal seams or breakdown of organic matter with limited gas or fluid escape through pore throats.

the products take up more space than the raw material from which they have been liberated. If the surrounding pore water does not escape, the fluid pressure increases as well. The same result occurs when natural gas, such as methane, migrates into an otherwise sealed unit from adjacent formations or proximal coal beds. The special requirement is that the gas move in at a considerably more rapid rate than the water is pushed out; otherwise, escape of the water relieves the pressure increase.

Osmosis, Hydrodynamics, and Abnormal Pressure

Osmosis is a process related to molecular diffusion. It essentially involves the movement of water molecules through a semipermeable membrane, from a less concentrated solution into a more concentrated solution, to eventually (by dilution) equalize the two concentrations. If the two solutions are salt water and fresh water, and the semipermeable membrane is a geologic barrier such as a shaley zone or a mylonitic, clay-filled fault plane, it might be argued that conditions favorable to osmosis could exist in subsurface formations.

In attempting to explain subsurface fluid phenomena, several aspects of the osmosis mechanism are particularly attractive. One problem is whether shales, clayey sediments, fault zones, and other natural subsurface features actually function as semipermeable membranes, by passing the solvent (water) from the fresh side and holding back the salts on the saline side. As Table 6.1 indicates, geologically significant osmotic pressure differentials across a barrier can be created within formation waters that are well within the range of those encountered in the subsurface (10,000–200,000 ppm sodium chloride [NaCl]; sea water is around 19,000 ppm NaCl). The few actual studies available do not support the possibility that this mechanism is a major factor in the development of overpressured reservoir units. This is unfortunate because the osmosis phenomenon explains some subsurface problems better than other physically feasible processes.

Figure 6.8 illustrates the hypothetical effects of osmosis on an oil-bearing reservoir in which a shaley zone performs as the semipermeable membrane. In this model, comparatively fresh water on the updip side of the barrier moves through it to equalize the chemical potential (electron quantity?) of the solutions on both sides and diffuses into the brine on the downdip side. The added water molecules increase the volume of the salt water and if, as illustrated in the figure, the pore fluids are physically confined, the internal pressure rises proportionally. The force of water expansion compresses the associated oil. The osmotic pressure increases until the resistance to it by the reservoir boundaries (back pressure) equals or exceeds it. At this point, the osmotic flow of the water across the barrier ceases. For the period of time that the process is operating, the freshwater side of the barrier and the barrier itself are hydrodynamic with water moving through the rocks, and the saltwater side (if sufficiently sealed) is (until the seals are breached) environmentally hydrostatic but overpressured.

It appears that underpressured reservoirs might be created by osmosis as well, as shown in Figure 6.9. In this case, the reservoir water is less concentrated than that on the other side of the semipermeable shale

TABLE 6.1. Theoretical pressure differences produced by osmosis across a semipermeable membrane separating solutions of different brine concentration.

		Salinity of the more dilute (fresher) formation brine		
	NaCl concentration	0 ppm	50,000 ppm	100,000 ppm
Salinity of the more concentrated (saltier) formation brine	100,000 ppm	(8.825 mPa) 1,280 psi	(4.688 mPa) 680 psi	0 psi
	150,000 ppm	(14.065 mPa) 2,040 psi	(9.928 mPa) 1,440 psi	(5.240 mPa) 760 psi
	200,000 ppm	(20.41 mPa) 2,960 psi	(16.27 mPa) 2,360 psi	(11.58 mPa) 1,680 psi

This table was constructed from values read off the graph presented by Bradley (1975) after the Petroleum Research Corporation, 1958.

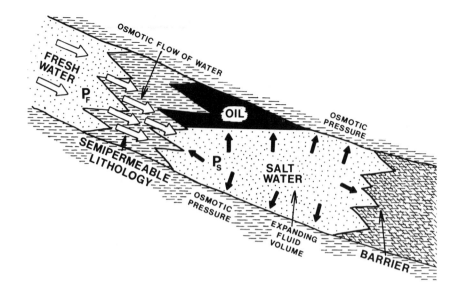

FIGURE 6.8. Osmosis can in theory create significant overpressures if the fluid on the more concentrated side of the semipermeable lithology is suitably confined so that as the volume of water expands and is unable to escape the pore pressure P_S rises.

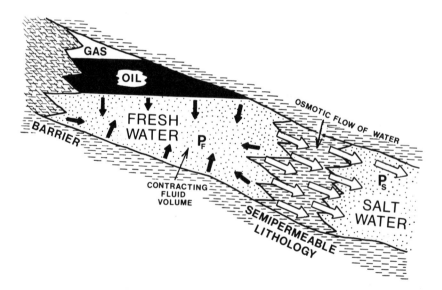

FIGURE 6.9. Underpressured reservoirs may likewise be produced by osmosis in a situation where the water underlying the hydrocarbons is significantly fresher than that on the other side of the lithologic unit functioning as the semipermeable membrane. The loss of water volume is compensated for by expanding hydrocarbons and the pore pressure P_F is diminished.

barrier. Suction of fresh water through the shale with no replacement from the surface would decrease the volume of water on the fresh side, the oil would expand to take up the increased space. Gas would come out of the solution, thus expanding the gas cap. The reservoir would become subnormal as the process continued which it would until the back pressure within the saline chamber into which the freshwater molecules were moving came to equilibrium with the osmotic pressure.

Once again, as attractive as osmosis is for explaining such phenomena as downdip flow, updip flow, geopressured reservoirs, and subnormally pressured reservoirs (as well as anomalous subsurface occurrences of fresh and saline formation waters), present evidence is woefully lacking. Future geologists may discover that similar but presently unknown processes have produced much of what puzzles us in the subsurface today, when we apply over-simplified and inappropriate concepts to explain such phenomena.

Geopressures

Occurrences of stratigraphically and geographically anomalous pressures have been described by researchers worldwide and numerous theories exist to explain how these conditions originate and are preserved. The most difficult aspect in understanding high-pressured units is determining just what kind of natural geologic barriers can maintain the very substantial pressure differences that have been documented by actual formation pressure measurements.

Vertical containment is created by impermeable units such as tight shales, anhydrite formations, and thin mineralized zones, which aren't always conformable with local bedding. The pore spaces within these zones display, by calcium carbonate or silica plugging, the spontaneous precipitation that might be triggered by a sudden loss of carbon dioxide from the system at some time during its postdepositional history.

Pressure-competent containment from the sides is considerably more puzzling. Mylonitized or mineralized vertical fault and fracture zones acting like the walls of a room are probably the major features responsible for pressure maintenance. Most likely, subsurface pressures are changing continuously, so that fluids are moving, the chambers are quietly leaking, and the entire subsurface environment is technically dynamic with respect to geologic time. In the practical sense with respect to reservoir production time, however, geopressured units are undoubtedly pretty much static and significantly different across the natural barriers.

Figure 6.10 illustrates a simple model that explains the pressure–depth gradient configurations typical of geopressured geologic sections. The model shows a set of "nested" grain-and-water-filled chambers, like glasses sitting inside each other. In the topmost chamber, pore pressures measured at AA and BB are on the same static gradient as those measured at A and B in the reference column. In the second chamber down, although the pressures measured at CC and DD are on the same static gradient, the gradient itself is displaced toward the higher pressure direction from the gradient upon which lie pressures measured at points A, B, C, and D in the uniform column. The difference is due to the weight of the grains and fluid in the overlying chamber (the weight is being applied to the grains and fluid in this chamber through the bottom of the top container). The bottom functions as a weakly supported seal in the model, and the degree of gradient offset is a function of the strength of the subseal grain framework. In other words, the grains and fluids in each chamber below the topmost have squeezing upon them the weight of everything above them. This weight is an external force pressing on the fluids, and the rest is Pascal's principle in action.

As an example of this type of gradient configuration, Figure 6.11 illustrates the pressure–depth relationships among the measurements from three wells in the Venture field on the Scotia shelf. The offset gradient segments are interpreted as reflecting separate pressure compartments bounded by barriers. As the figure indicates, there are 1,000-psi differences between the individual systems. This magnitude seems

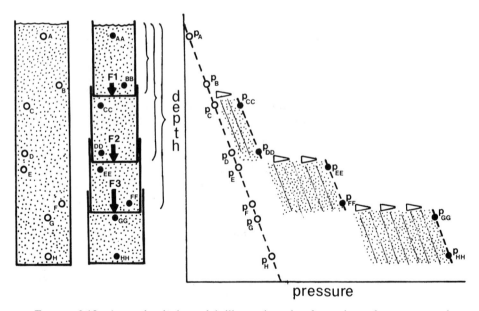

FIGURE 6.10. A mechanical model illustrating the formation of geopressured sequences and compartments reflected by discontinuities in the pressure–depth regional gradient. Individual within-compartment gradients are presumed to be close to static gradient for the fluids concerned.

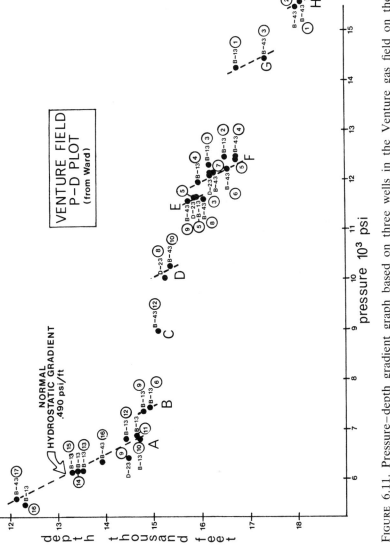

FIGURE 6.11. Pressure–depth gradient graph based on three wells in the Venture gas field on the Canadian Scotia Shelf. (Courtesy of G. Ward.)

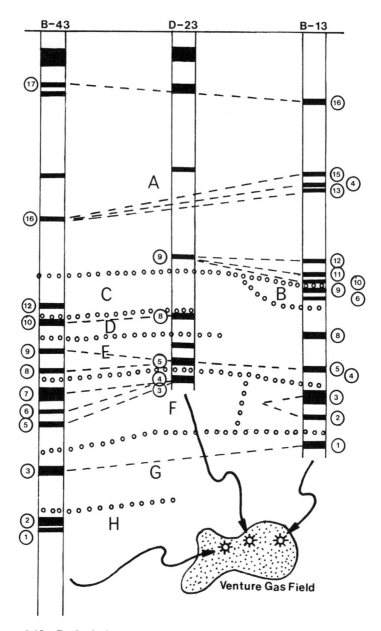

FIGURE 6.12. Geologic interpretation of correlated pressure zones in the three wells of Figure 6.11. Compartment-like geometry is evident. (Courtesy of G. Ward.)

much too high to account for by hydrodynamic flow. The flow pattern would be almost unrealistically complex, particularly in light of the stratigraphic pressure system relationships illustrated in Figure 6.12.

There are both favorable and unfavorable aspects of geopressured reservoirs. On the plus side, they can hold more gas in place because of natural compression, and rock pores can be held open, resulting in anomalously high porosities for given depths compared with normally compacted sediments. Because of this, lithologically non-cost-effective reservoirs can be produced economically. On the negative side, over-pressured reservoirs present technical problems, making them much more expensive to produce. As well, failure to predict and detect overpressured zones during drilling can result in unwelcome "kicks" (gas invading the well bore) and even blowouts and leaks. At the same time, densities and sonic velocities in overpressured units are often anomalous, so that seismic interpretation is strongly affected by the development of abnormally high pore-pressure-induced porosity in isolated zones at unexpected depths.

7

Pressure Variation with Depth

Pressure–Depth Gradient Diagrams

Useful information reflecting unknown attributes of subsurface oil, gas, and water systems can be gained by constructing and interpreting scenarios from simple graphical plots of formation pressure against the depths (or elevation) at which they are measured. There are essentially three exploration problem areas in which such P–D plots are applied: the first concerns the prediction of pay thickness (i.e., placing oil–water, gas–water, and gas–oil interfaces); the second concerns subsurface correlation and the prediction of prospective reservoir trends; the third involves the recognition of subsurface barriers and the definition of geologic pressure systems and placement of boundaries with respect to over- and underpressuring.

P–D gradient analysis is based on the principle that the static gradient for a fluid reflects only the density of the fluid concerned so that deviations from it must indicate additional influences, presumably geologic. Figure 7.1 shows simple static gradients for six common fluids corresponding to the tanks described in chapter 1 (Figure 1.4). The slopes of the gradients (rate of increase in pressure with increase in depth) mirror the densities of the individual fluids (e.g., 0.050 psi/ft for the natural gas, compared with 0.493 psi/ft for the saline water), and since the tops of the columns have been pressurized to 1,000 psi, the lines demonstrate graphically the rate at which the pressures increase from 1,000 psi down. If pressure gauges were inserted through the walls of the tanks at varying depths, the observed pressures would lie on the gradient lines. These fluid systems are hydrostatic and at rest.

If a single large tank is filled with several immiscible fluids of different density, the lighter ones will float on the heavier ones, as illustrated in Figure 7.2, and the pressure will vary within each fluid layer according to its density. More important, the depths at which the gradients intersect each other will reflect the positions of the transition zones between the individual layers. The important point is that if one could not see into the tank but had the means to measure pressures through its walls at various

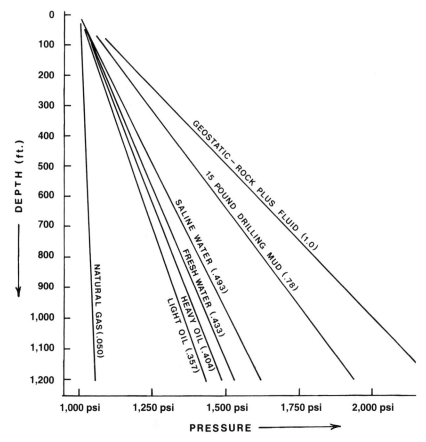

FIGURE 7.1. Pressure–depth (or pressure–elevation if corrected to sea level) gradient plot showing simple static gradients for six common subsurface fluids.

depths, the gradients plotted on a P–D graph, as shown, would indicate the density of the fluids concerned, and the gradient intersections would reflect the positions of the contacts between each of the layers. In this way, within a subsurface reservoir the relative proportions of oil, gas, and water can be predicted from pressure values.

Geologic Interpretation of P–D Plots

Formation pressure data reveal aspects of subsurface reservoir complexes that cannot be discerned using information from other sources. Gradients representing individual systems can be established on P–D graphs when (1) two or more pressure points can be connected by a straight line, the

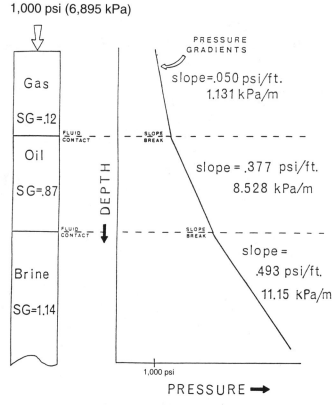

FIGURE 7.2. Static gradients in a three-phase system illustrating the relationship between the fluid interfaces and gradient slope breaks.

slope of which indicates the density of the fluid concerned, and (2) a single pressure point plotted at a particular depth through which a line with a slope corresponding to the known or assumed density of the fluid, is projected.

A practical example of P–D gradient analysis is shown in Figure 7.3. The first well drilled to test a subunconformity sand in this area is well A. This well is bottom-hole tested, yielding pressure P_1 and a significant amount of formation water. P_1 is plotted on the P–D plot, establishing a reference point for that fluid system through which an appropriate brine gradient line is projected. The slope of this line is based on the density of the recovered water, either from a chemical analysis or assumed. It is apparent to the sponsor of this prospect that the initial well is too low, so well B is proposed at a location higher to the east. This well enters an identical water-wet sand, which is tested. When plotted on the graph, P_2 falls a significant distance off the original gradient established by P_1, indicating that as a member of a different, higher-pressured system this sand has to be a different one from that found by well A—so the prospect is still alive. A very persuasive promoter might succeed in getting well C drilled between well A and well B. Well C verifies that there are indeed two separate sands. The bottom-hole test yields water and pressure P_3, which when plotted on the graph falls on the same gradient as P_2, indicating that they have indeed tapped the same reservoir. Two pressures are assumed to be related system-wise when they lie close to the same density gradient. Generally, though not always, this is a safe bet. P_4 is measured in a test that recovers significant gas, and it plots at the location indicated in the figure. Projecting a gradient with the appropriate gas-density-related slope down through this point to where it intersects the initial water gradient brackets the vertical dimension of the gas column. The presumed gas–water contact can be picked at the depth of intersection of the water gradient and this gas gradient. If there is an oil leg (as shown), the gradient for it would intersect the others at the points indicated in the figure. Of course, this would be speculative only, since none of the wells in the example have penetrated oil and none of the tests have recovered any.

The hydrocarbon pay thickness estimates yielded by this procedure can be applied to calculating reserves and evaluating economics for this prospect. There is no other way for predicting hydrocarbon–water contacts that have not actually been drilled.

The three parameters interconnected through physics in such a way that any two of them determine the third are, as shown in Figure 7.4,

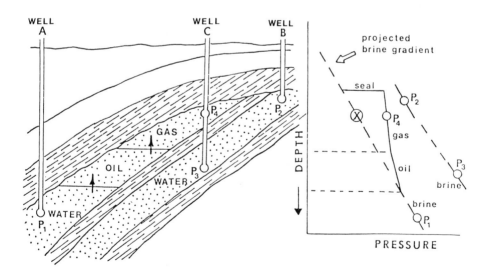

FIGURE 7.3. Pressure–depth gradient plot that reflects the geologic scenario and provides answers to important exploration and production questions that could not be answered any other way.

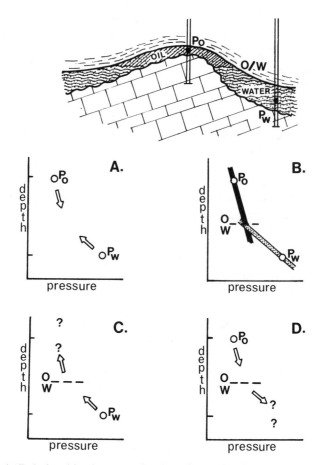

FIGURE 7.4. Relationships between the three interrelated parameters of oil pressure, oil–water transition zone, and pressure in the underlying water leg.

pressure at any location in the hydrocarbon leg (oil or gas), position of the hydrocarbon–water contact, and pressure at any location in the underlying water leg. From the two pressures, the contact can be positioned (A) and (B). From an underlying water pressure and the known or assumed position of the contact, the pressure at any point in the overlying oil reservoir can be determined for production, completion, or reserves estimation purposes. And with a reservoir pressure and knowledge of the position of the oil–water or gas–water contact, the pressure anywhere in the associated water leg can be predicted and used for calculating the potentiometric surface elevation, recognizing overpressured formations or projecting oil–water contact positions in related reservoirs.

A portion of a subsurface structure is shown in Figures 7.5, 7.6, and 7.7; it is gas and oil bearing with brine underneath. In Figure 7.5, two

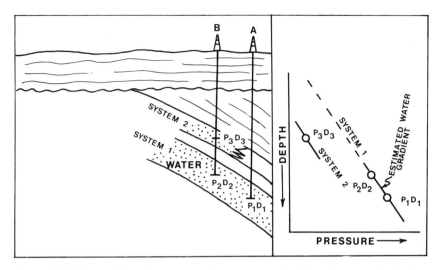

FIGURE 7.5. Two wells and corresponding pressure–depth gradient graph showing inferred water systems.

wells have been drilled. After the first, with the recovery of formation water the density of which was 65 lb/ft^3, it was possible to plot the density gradient through point (P_1D_1) on the accompanying P–D graph. The second well (P_2D_2) was drilled updip but not quite far enough on the inferred structure, and it plotted close to the previously projected gradient

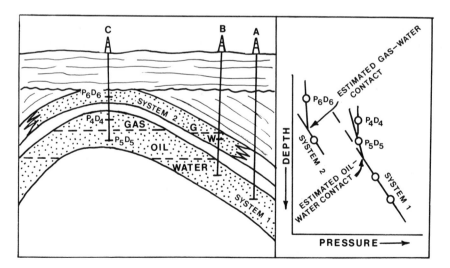

FIGURE 7.6. Three wells and corresponding pressure–depth gradient graphs showing inferred water systems and gas and oil legs.

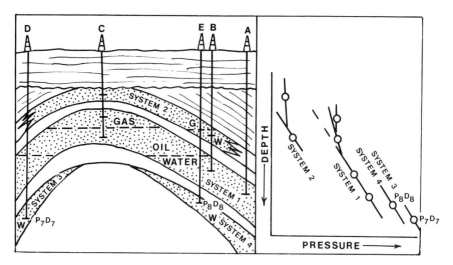

FIGURE 7.7. Five wells and corresponding pressure–depth gradient graph showing inferred hydrocarbon and water systems.

for the system established from (P_1D_1), verifying that the same water system had been tapped. (P_3D_3), another water test, lies well off the initial gradient, suggesting that it represents a different system (which in fact it does). Well C, drilled on the crest of the structure, encounters gas and oil and yields pressures P_4D_4 and P_5D_5 measured in the gas and oil legs of the lower sand. These points appear on the P–D graph in Figure 7.6. An oil-density gradient projected downward from (P_5D_5) to the point of intersection with the original water gradient indicates the thickness of the oil column. Plotting point (P_6D_6) at its proper position on the graph and projecting a gas gradient from it down to where it intersects the System 2 water gradient shows that a separate gas reservoir with a considerable gas column is likely.

Well D, drilled to test the other side of the structure (Figure 7.7), verifies the continuation of the oil reservoir and the termination of the overlying gas reservoir. It likewise encounters some deeper sands, which are water wet and may be correlated with those penetrated by well E back on the east side of the structure. The gradients through points (P_7D_7) and (P_8D_8) may really not be significantly different from each other. This becomes a matter of interpretation.

Regional Systems: Separate Hydrocarbon Reservoirs on a Common Aquifer

Figure 7.8 shows a case from central Alberta in which a series of Devonian reef oil pools, although physically and geographically separate from each other, are connected by a common water system or aquifer. The different

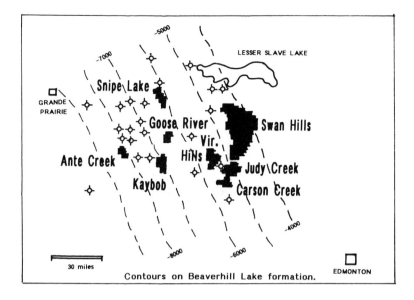

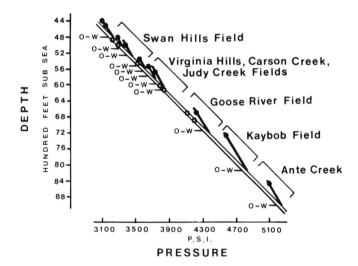

FIGURE 7.8. Example of a regional hydrogeologic system from central Alberta showing Devonian Beaverhill Lake oil reservoirs, which are connected by a common underlying aquifer, the Gilwood sand. (From Hemphill et al. 1970.)

oil–water contacts for the individual pools are positioned on the P–D graph in the figure, based on early drilling results and projected oil-density gradients. The Homeglen-Rimbey Trend in Alberta is another example of a chain of individual reefs in fluid communication with each other via a common underlying aquifer, as illustrated in the cross section

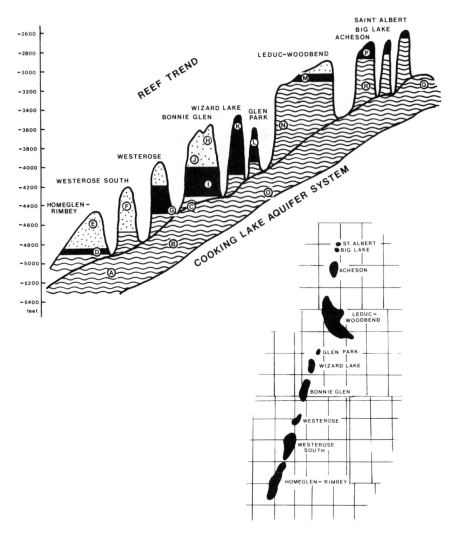

FIGURE 7.9. The Homeglen-Rimbey Trend in Alberta, with a string of single oil- and gas-filled reef reservoirs that are in fluid communication with each other through the underlying Cooking Lake formation aquifer. (From Oilweek Magazine, 1984.)

in Figure 7.9. Anomalous pressures in a largely open geologic system such as this are therefore only transitory, since sudden changes due to production removal of hydrocarbons are compensated for quickly by the regional waters.

A problem concerning this type of regional system with a series of separate pools related to a common aquifer is illustrated in the geologic cross section in Figure 7.10, with the accompanying empty P–D plot to be completed by the reader as a practice problem. No numbers are involved in this problem, merely the P–D relationships as they can be discerned from the cross section. The pressure measured at (A) is the entry point to the graph; from point (A) the appropriate water gradient can be projected upward at an angle matching the water gradient slope shown in the figure. The other tests in the formation water phase of the aquifer can be added to the graph in the proper positions by following the dotted elevation lines to the points of intersection with the water gradient. The succeeding oil and gas pressure points can be located once the regional water gradient is established by projecting oil or gas gradients with slopes as provided in the figure, upward from the elevations of the fluid contacts shown in the cross section. The oil and gas test gradients intersect the water gradient at the depth position of the contact. By using these relationships and the key gradient slopes, the P–D plot can be completed to reflect the hydrogeology. Pressure measurement (B) should be posted on the graph in its indicated position, showing that the stratigraphically younger reservoir that it comes from is a system that is independent from the others. (The correct answer is in the Appendix.)

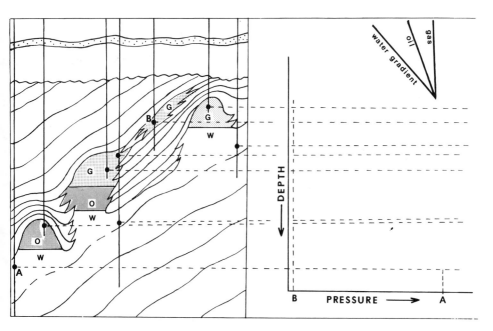

FIGURE 7.10. An illustrative problem to be completed by the reader. The proper P–D gradient plot should be constructed to reflect the geologic and fluid relationships shown in the cross section.

Structural Trap Reserves Estimation Problem

Seismic surveys suggest the sizable structure shown in Figure 7.11. A well drilled at location A tested brine at $-1,840$ ft (-560 m), the formation pressure was 3,150 psi. The salinity of the water is 210,000 ppm TDS. Assume that the structure contains gas and the pressure in the gas at its crest (-950 ft or -290 m) is about 3,000 psi, with a static gas gradient of 0.090 psi/ft.

The problem is to predict just how full this structure is likely to be within the limits of these parameters. Construct the P–D gradient diagram to predict the elevation of the trace of the anticipated gas–water contact. Since the fluid environment is hydrostatic and the gas–water contact is horizontal, it will parallel the structure contours. Is the structure likely to be close to full? or empty?

If further analysis of the seismic data suggests an oil–water contact at $-1,800$ ft, and the oil is assumed to be 35° API gravity, add that gradient to the graph and map the trace of the gas–oil contact. How many feet of oil and gas pay might then be anticipated in this structure? Use Table 1.2 to find the proper gradients, or calculate them yourself. The gradients can be drawn on the graph in Figure 7.12 by figuring pressure differences over

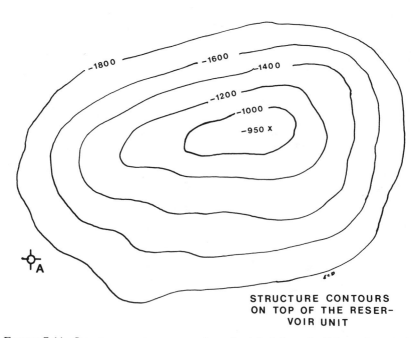

STRUCTURE CONTOURS
ON TOP OF THE RESER-
VOIR UNIT

FIGURE 7.11. Structure contour map of a seismicly inferred offshore feature for predicting trap capacity using pressure–depth relationships.

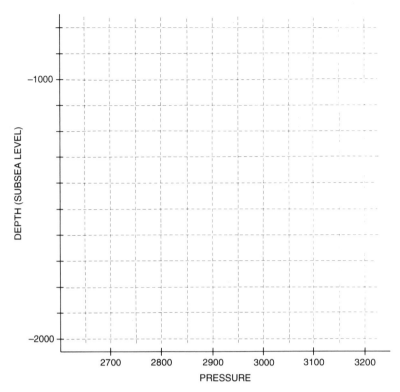

FIGURE 7.12. Empty graph for solving the trap capacity problem described in Figure 7.11.

an interval of 1,000 ft, making a little construction point on the graph and then connecting the actual pressure point and the projected construction point with a straight line. For example, if formation pressure in a fresh-water reservoir at $-1,500$ ft is 2,100 psi, at -500 ft (1,000 ft above) it is 1,200 psi $-$ 433 psi $=$ 767 psi. The line connecting the first point and the second represents the gradient on the graph.

The relationship among the position of the oil–water contact, the oil-density gradient, and the water-density gradient is illustrated in Figure 7.13, which is a schematic P–D gradient plot. Also shown are changing oil pressures in a roofed reservoir. The pressure in the underlying water is assumed to be constant and is represented in the figure as P_W. Three hypothetical pressures in the oil column are shown (A, B, and C), with appropriate oleostatic (oil-density) gradients. As the oil pressures increase from A to B to C along the pressure axis, the points of intersection between the oleostatic and hydrostatic gradients move down to lower and lower depths as the oil displaces the water. The corresponding oil columns

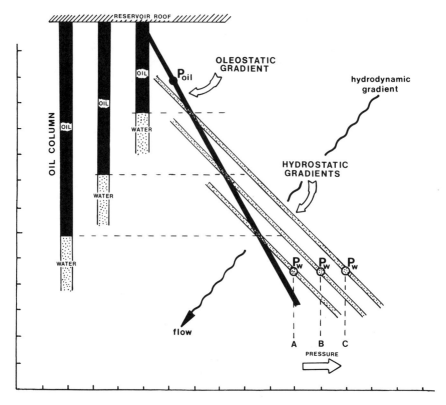

FIGURE 7.13. Conceptual relationship between sealed reservoir oil column and position of the oil–water transition zone as reflected by changing (density and load) magnitudes of water pressure.

implied by the gradient intersection positions are shown for comparison. The buoyancy pressures at any depth are equal to the difference between the pressure in the sealed reservoir at the depth and the corresponding hydrostatic pressure (estimated from the water gradient) at the same depth, as shown in the figure. At the very top of the reservoir this is essentially the entry pressure that must be *exceeded* by buoyancy to move oil into the top seal.

At each particular pressure within the oil leg, the height of the oil column is also affected by the pressure within the underlying water leg, as indicated in Figure 7.14. The greater the internal pressure at some depth in the water leg, the thinner the oil column, as illustrated. At the lower pressures, the oil column is the thickest and buoyancy pressure of the oil at the top of the reservoir is the greatest. The displacement pressure in the seal is correspondingly high, and the seal is competent enough to hold back the hydrocarbons.

An oil-bearing sand pinchout is shown on the map in Figure 7.15. One well has been completed as an oil well; when originally tested, the shut-in pressure was 2,880 psi measured at a subsea elevation of −3,836 ft. This test likewise recovered 600 ft of oil. Downdip from this location at a depth of −4,002 ft, the pressure in the underlying water leg is 2,940 psi, and the water seems quite fresh. At the same time, the desirability of the three land blocks depends on whether the most likely location of the oil–water contact is downdip close to the water well, in which case these blocks are attractive to investors, or updip near the productive well, indicating that the parcels are more likely to overlie water rather than oil. To evaluate the situation, construct the P–D gradient graph to determine the position of the hydrostatic–oleostatic gradient intersection, which is the oil–water contact. (This can be done on the empty graph in Figure 7.16.)

Another similar problem requiring estimation of the positions of fluid contacts in a unit on a considerably greater geographic scale is presented next. (This problem has been contributed by Mr. Fred Meissner of Denver, Colorado.) An exploratory well drilled at a location south of the

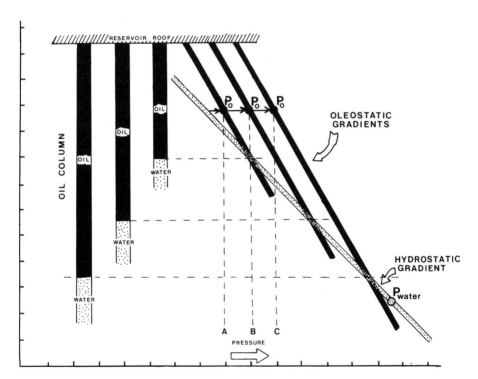

FIGURE 7.14. Conceptual relationship between sealed reservoir oil column and position of the oil–water transition zone as reflected by changing (buoyancy) pressure levels in the oil zone.

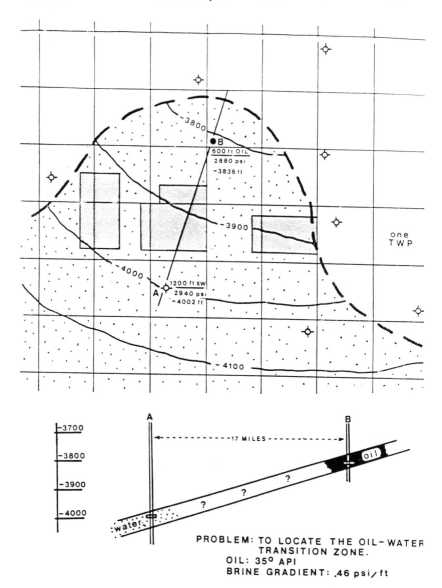

FIGURE 7.15. The reader is invited to construct an appropriate P–D graph and use it to evaluate the acreage blocks with respect to the most likely position of the oil–water contact.

Zagros Mountains in Iran (see Figure 7.17) tests "high pressure gas with a substantial condensate content" in the Asmari limestone in the subsurface of Iran. The best estimate of the formation pressure in the gas leg is 4,995 psig at a subsea depth of $-7,363$ ft. The density of the gas, corrected to reservoir depth, is 0.35 gm/cc. (To express the appropriate static gradient in SI units, use Table 1.1 to convert gm/cc to kg/m^3 and SG to grad P.)

Recognizing this as a possible gas cap on a larger downdip oil accumulation, a second well is drilled 17 mi downdip from the first. This

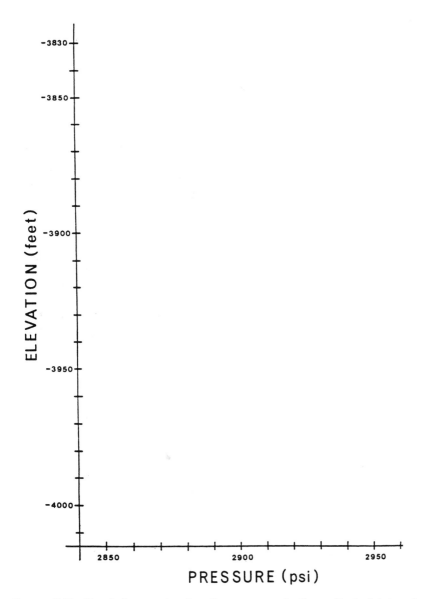

FIGURE 7.16. Graph for constructing the pressure–depth gradient plot to solve the problem described in Figure 7.15.

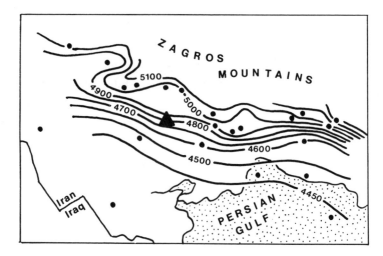

FIGURE 7.17. Map of the Marun oil field area in Iran (from Fred Meissner, Denver, Colorado). The reader is invited to predict the oil pay thickness by constructing the appropriate pressure–depth gradient plot. The triangle shows the location at which the Asmari formation occurs at a depth of 9,000 ft.

well finds the top of the Asmari at −9,549 ft and the bottom of the unit at −10,722 ft; this entire interval is oil saturated, and the density of the oil under reservoir conditions is 0.62 gm/cc with a measured fluid pressure of 5,731 psig at a depth of −10,534 ft. (What is the static gradient for this oil?) These data can be used to predict the positions of the gas–oil and oil–water contacts even though neither has been penetrated by the drill, provided that one additional piece of information on the character of the bottom water is available.

From regional hydrologic pressure data, it is estimated that at −9,000 ft, the formation pressure in a water-saturated Asmari section should be about 4,800 psi. Using a water density of 1.0 gm/cc, this estimated pressure point can be established on the P–D graph, and a "regional" water gradient can be projected through it. The intersections of the three gradients plotted through the gas point, the oil point, and the water point yield a P–D picture indicating how large the oil pool is likely to be under these conditions if it is continuous. Use the graph in Figure 7.18 to solve this problem.

Subsurface Correlations Based on Pressure Data

The recognition of reservoir continuity from a common pressure system is accomplished using pressure–depth data to identify groups of wells that are members of the same or different pressure systems. This procedure is extremely helpful in correlating geologic units between widely scattered wells to predict the extent of the particular reservoir formation concerned.

The concept is illustrated in Figure 7.19, which shows an area within which four wells have been drilled; three of the wells are gas producers and one has recovered water in a 50-ft-thick sand unit. Of interest are the prospects within the blocks in the central portion of the area. Several possibilities exist. If (3) and (4) are in the same single reservoir unit, then the blocks along the trend between them are favorable for evaluation drilling. If (1) and (4) are producing from the same single gas sand, then the blocks between their two locations are highly prospective for more production. It is also possible that each single well is tapping an isolated reservoir, which, if true, means that none of the central blocks can be recommended; on the other hand, there is still the chance that all the wells are producing from the same uniform reservoir that extends across the entire area.

Because of long distances among individual wells, the continuity of a reservoir unit often cannot be established by geologic criteria, even though the well logs display strong lithologic similarities. The objective is to discern whether there is likely to be a single continuous reservoir, as

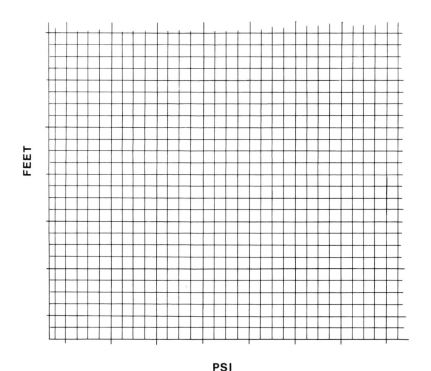

FEET

PSI

FIGURE 7.18. Graph for the Marun oil field problem.

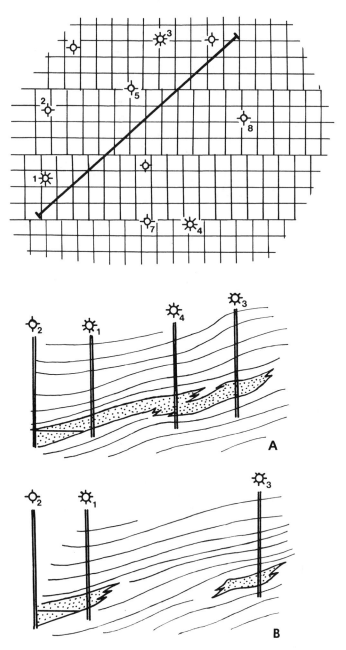

FIGURE 7.19. Two possibile reservoir scenarios for a gas sand penetrated by four widely scattered wells.

suggested in cross section A, or several separate sand bodies, as illustrated in cross section B.

When the pressures are plotted on a graph, such as in Figure 7.20, and the points appear to lie close to the same single gradient, the implication is that the pressures are from widespread locations within the same single large reservoir. This result is not completely conclusive, since it is possible for pressures from physically isolated reservoirs to lie near the same gradient as members of the same pressure system. Pressures can persist through less permeable rocks that physically seal off hydrocarbon reservoirs.

A significant footnote to this discussion is that this concept was the tip-off behind the discovery of western Canada's one trillion cubic foot Ring-Border gas field in 1988. As described by Masters (1991), the reservoir is in a Triassic sand "wedged beneath an unconformity and covered by Cretaceous shale and other sands." Five producing but noncommercial gas wells had tested more than 5 mmcf/d, and the government of British Columbia had decided that each was tapping a single pool or reservoir only. A simple P–D plot would have shown otherwise, though, since eventual studies by the Canadian Hunter Oil Company disclosed that the pressures in each of these wells plus eighteen other "dry holes" were related to a single fluid column. This demonstrated that the reservoir was more likely to be a single blanket accumulation than a collection of "scattered ponds" with plenty of available acreage. This result led to the recognition of the gas pool, a real-life case of the model scenario illustrated in Figure 7.20.

If, on the other hand, the gas pressures when plotted on the graph are scattered such as in Figure 7.21, then it is apparent that (assuming the

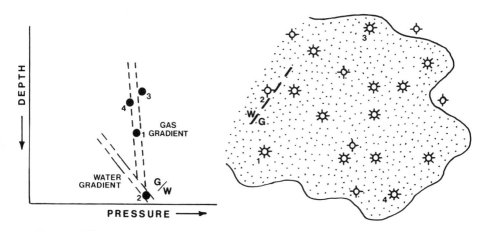

FIGURE 7.20. A single-system interpretation for the gas sand reservoir inferred from the P–D gradient configuration (there is a roughly 60% possibility of this interpretation being correct).

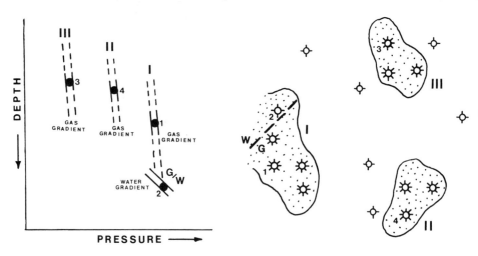

FIGURE 7.21. A multisystem interpretation for the gas sand reservoir inferred from the P–D gradient plot (it is extremely likely that this interpretation is correct).

pressures are accurate) several different pressure systems are reflected, each typified by its own gradient, as indicated in the figure; pressures from subsequent wells will cluster close to these gradients. In this case, the geologist can be reasonably certain that different systems (and thus separated physical reservoirs) are represented, each with its own unique gradient.

The results of this procedure can be applied in mapping reservoir trends to serve as a basis for lease acquisition and drilling, as illustrated in Figure 7.22. The P–D plot suggests the possibility of at least three separate systems, which define shaded trends on the map. Favorable acreage is indicated by the blocks in the figure. Numerous companies rely on this procedure to lead them to lucrative gas reserves in Alberta.

Putnam and Oliver (1980) used pressures for predicting channel sand trends in the Lower Cretaceous of Alberta. A base map showing the location of wells, many of which were completed as gas wells, is shown in Figure 7.23. The available gas pressures for these wells from production or DST data are listed in Table 7.1. These data can be used to construct a P–D gradient plot to ascertain the number of possible systems and identify which wells appear to have tapped which system(s). It is necessary to post the well label by each of the points on the P–D plot, so that the proper system can be assigned to the correct location on the map in Figure 7.23. When the systems are identified using all the possible wells, the trends can be sketched (as in Figure 7.22) and the prospective blocks of acreage can be duly highlighted. (Use Figure 7.24 to construct the P–D gradient plot.) The identification of these reservoir trends is the principal objective of this exercise. Sometimes, dry holes along the apparent trend can be wells in which gas production has been overlooked; in practice, these should be reexamined carefully.

We can speculate on natural mechanism(s) that might create a significant pressure difference in two adjacent sand channels with similar geologic histories in terms of overburden load, density of formation water, present-day depth, geothermal gradient, maturation history, and so on. One possibility is the comparative competence of the encasing lithology and resulting differential rates of fluid expulsion that occur in response to burial and overburden stress, as illustrated in Figure 7.25. If

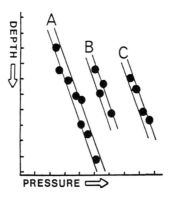

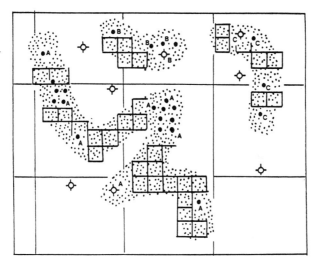

FIGURE 7.22. Inferred reservoir trends from gradient configuration showing parcels of "trend" acreage implied by families of wells that are members of the same system.

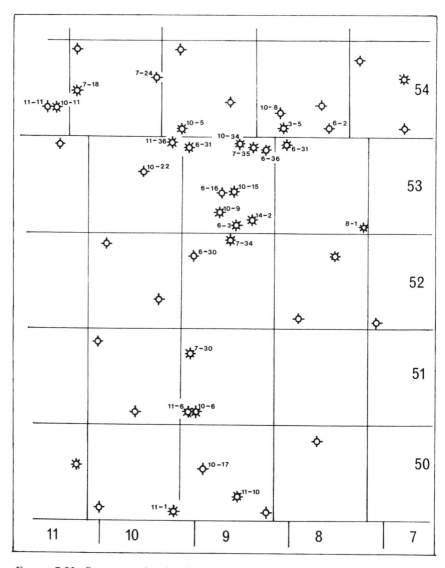

FIGURE 7.23. Base map showing locations of wells in area where gas production originates from Cretaceous sand channels. (From Putnam and Oliver, 1984.)

the fluids cannot escape easily (such as in channel A), then the rate of load-related pressure buildup over time is rapid. In channel B, the rate is slower, since some fluid escape occurs in response to the overburden load; pressure buildup and maintenance over time is moderate. The third channel, C, is almost an open system, so the degree of pressure increase above hydrostatic attributable to load-induced squeeze is relatively small; the pore fluid pressure in this well is close to normal. The pressures encountered in these three channels today would therefore be quite different even though the geology is the same.

Another explanation for short-distance pressure differences is migration into tightly sealed reservoirs of differing volumes of gas, either from proximal coal beds or from cooking of indigenous and varying amounts of organic matter. If the seals are strong, the gas is contained until its buoyancy pressure exceeds the displacement pressure of the surrounding rock. It would indeed be a coincidence if volume and pressure of the

TABLE 7.1. Cretaceous channel well data corresponding to the locations in Figure 7.23.

Well	PSI	Elevation
7-27-48-8	625	+245
11-10-50-9	612	+252
10-17-50-9	580	+255
11-1-50-10	635	+230
10-6-51-9	600	+290
11-6-51-9	610	+298
7-30-51-9	628	+275
6-30-52-9	590	+372
7-34-52-9	580	+330
"	610	+335
"	615	+305
8-1-53-8	530	+375
6-31-53-8	520	+452
14-2-53-9	600	+340
6-3-53-9	618	+350
10-9-53-9	608	+308
"	610	+320
"	595	+362
10-15-53-9	610	+370
6-16-53-9	580	+352
6-31-53-9	528	+405
10-34-53-9	535	+408
7-35-53-9	515	+410
6-36-53-9	520	+400
10-22-53-10	615	+322
11-36-53-10	602	+382
"	615	+352
6-2-54-8	505	+410
3-5-54-8	512	+450
10-8-54-8	510	+435
10-8-54-8	530	+450
10-8-54-8	532	+300
10-5-54-9	535	+398
7-24-54-10	542	+338
7-18-54-10	592	+345
10-11-54-11	605	+328
11-11-54-11	605	+348

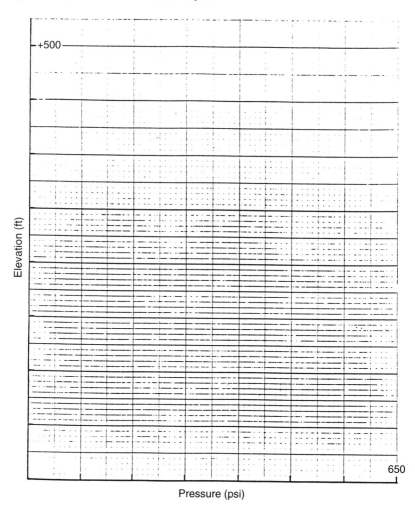

FIGURE 7.24. Graph for constructing pressure–depth gradient plot to solve channel problem in Figure 7.23. Appropriate gas gradient for testing point clusters is shown.

interstitial gas in one channel reservoir was the same as those for another. Sometimes, the higher the pressure, the smaller the reservoir, but not always.

Interpretation of Pressure Differences on P–D Graphs

Positions of points on P–D plots can reflect the local geology of the units that they represent. Some such relationships are illustrated in Figure 7.26. Top case shows two pressures measured at two different depths, P_1 and

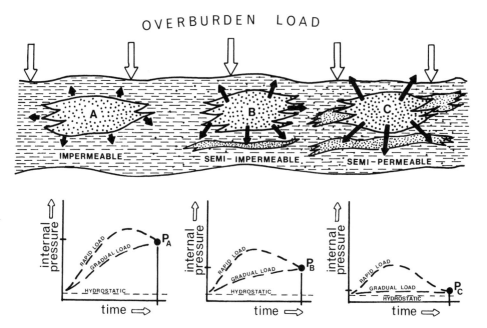

FIGURE 7.25. Differential fluid expulsion model to explain pressure differences among sand reservoirs that are geographically close to each other and have been subjected to the same historical factors.

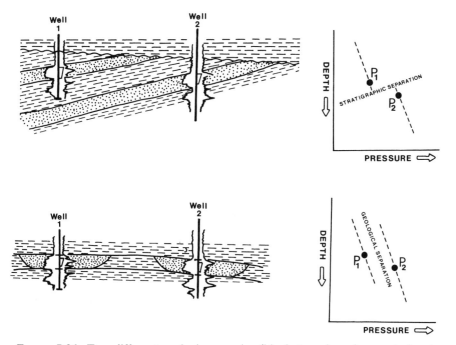

FIGURE 7.26. Two different geologic scenarios (blanket sands and separate bars/channels) to explain pressure differences on the pressure–depth plot.

P_2. Their positions on the graph may reflect two lithologically similar sands of different age that are not in the same regime and thus are separate systems (not on the same gradient). Or the points may represent pressures from two different geologic features, like the two channels of the same age shown in bottom case. In this illustration, the separation reflects the presence of a geologic barrier such as pressure-resistant fault plane, tight facies, or diagenetic zone.

Such a pressure point configuration can occur even when both units are measured in a single, homogeneous water-saturated unit, as illustrated by top case in Figure 7.27. This is produced by water flow from the vicinity of the higher-pressure well toward the lower-pressure well (actually, from high potential energy toward lower potential energy, but in this case the two agree). The formation here is hydrodynamic rather than static.

Finally, as illustrated by bottom case, there is the possibility of error in the pressure measurements, either one or both. The sand is hydrostatic with a single gradient, but either P_1 is erroneously high or P_2 is erroneously low (or both). These cases serve to illustrate just a few of the many interpretations from which a geologist must choose when working with P–D data.

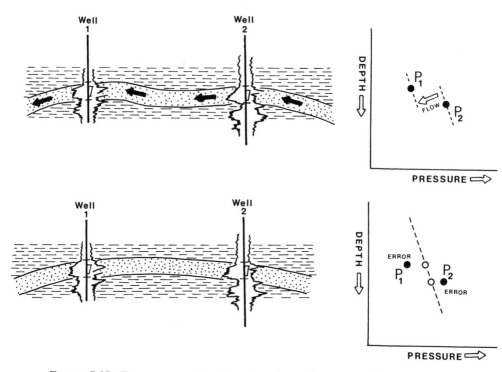

FIGURE 7.27. Two more explanations for observed pressure differences on the pressure–depth gradient plot.

P–D Gradient Plots and Potentiometric Surfaces

The pressure-dependent height to which a fluid will rise (independent of capillarity) in a manometer tube (or a well) can be estimated graphically from a P–D plot as well as calculated. When a particular gradient line, such as those in the P–D graph in Figure 7.28, is projected upward and to the left to a point corresponding to a pressure of 1 atm (14.7 psi) on the pressure axis, the elevation of this point indicates the hydraulic head calculated as $(H_W) = KB - RD + P/grad\ P$. This elevation physically corresponds to the theoretical position of the water-column–atmosphere interface. As the figure shows, the hydraulic heads calculated from the lower pressure points are the lowest, and those based on the higher pressures are the highest.

If the system represented is nonhydrodynamic, the points most likely reflect barrier-related pressures at different locations in separated cells or compartments, such as the fault blocks illustrated in Figure 7.29. The

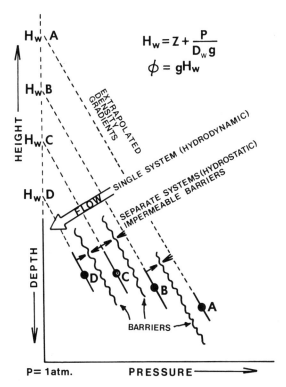

FIGURE 7.28. Pressure–depth gradient plot showing the relationship between pressure and hydraulic head and the inferences behind attributing observed pressure differences on the graph to either "flow" or "no flow."

trace of the potentiometric surface inferred from the hydraulic head elevation control points is stepped as shown, and the discontinuities coincide with the locations of the system boundary barriers. On the other hand, if familiarity with the geology dictates that the fluid system is most likely regionally continuous, as illustrated in Figure 7.30, then the trace of the potentiometric surface is more likely to be similarly continuous and tilted, reflecting a hydrodynamic unit. (In the first case, the possibilities for entrapment are more numerous than in the second!)

Pressure-gradient graphs are helpful aids to subsurface geologists; however, useful interpretation of the pressure relationships is enhanced by a sound understanding of the geologic framework as well. In addition, formation pressures are often highly inaccurate, so potentiometric surface differences of less than 50 ft may not be meaningful because of the natural scatter of the data (statistical accidents). It is sometimes more appropriate to map hydraulic head ranges as "bands" that reflect generalized "regions" rather than contour gradients. The group boundaries can be set up empirically from the P–D diagram, as illustrated in Figure 7.31, or calculated to yield hydraulic head threshold values that correspond to the pressure breaks. The H_W threshold values are calculated by substituting

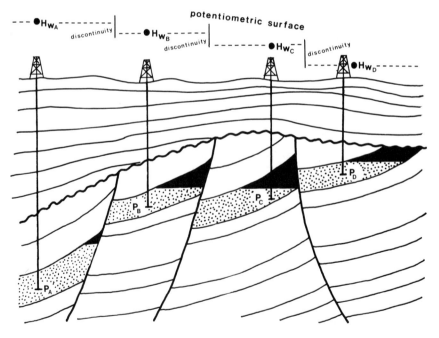

FIGURE 7.29. Hydraulic head elevations in cross section inferred from the pressure–depth relationships in Figure 7.28 and the assumption of "no flow" geology.

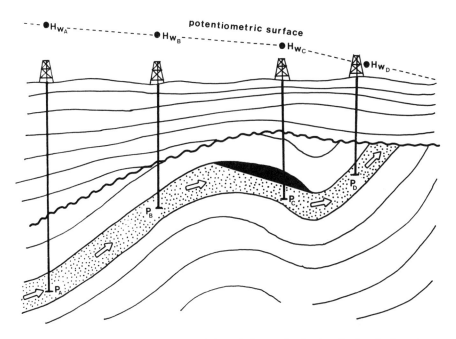

FIGURE 7.30. Hydrodynamic flow inferred from the hydraulic head elevations calculated from the pressure–depth data in Figure 7.28.

the appropriate P, grad P, and Z (baseline) values into the equation H_W = Z + P/grad P. In the example, if the pressure cutoffs P_1 and P_2 are 2,000 psi and 3,000 psi at a baseline elevation of −3,000 ft, and grad P is 0.46 psi/ft, the corresponding H_W cutoff elevations will be 1,438 and 3,522 ft *above* sea level. A map of the type illustrated in Figure 7.32 can be constructed using different patterns for the areas typified by each particular level of hydraulic head. The comparison of these areas with the interpreted structure suggests the possibilities of major fault traces functioning as pressure barriers, as shown, presenting the possibilities of several potentially prospective hydrocarbon-bearing traps formed by the fault–barrier intersections. This interpretation fits compatibly with the existing producing wells and extends the size of the individual pools significantly.

Pressure Deflection Maps

A similar method for presenting hydrologic data is by constructing a map that relates the various systems to one reference gradient, which becomes the standard. A scale line (corresponding to what we have referred to as the "potentiometric gradient" or "flow gradient") is constructed *normal*

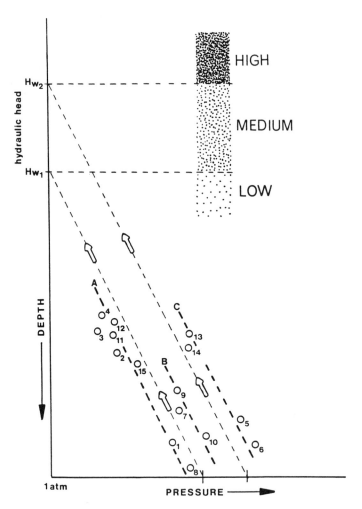

FIGURE 7.31. Inference of separate systems from clusters of points on a pressure–depth gradient plot.

to the local hydrostatic gradient, and values are assigned to the original pressure points based on their position along this line. The standard system is defined as zero with numbers on the high potential side positive and those on the low side negative. The score, therefore, indicates the extent that a given point departs from the standard.

These values are then plotted on the map at the well locations, and even contoured. The result is a type of coded potentiometric map, showing departure values rather than hydraulic head elevations.

It should be kept in mind that pressure variations within or between systems on the P–D graph are measured in the horizontal direction,

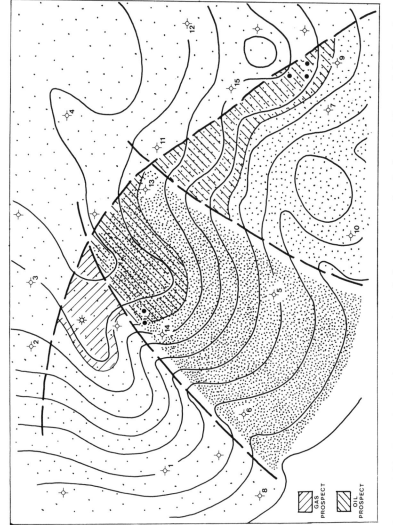

FIGURE 7.32. Fault barrier interpretation of the pressure–depth data in Figure 7.31 showing prospect south of Well 11.

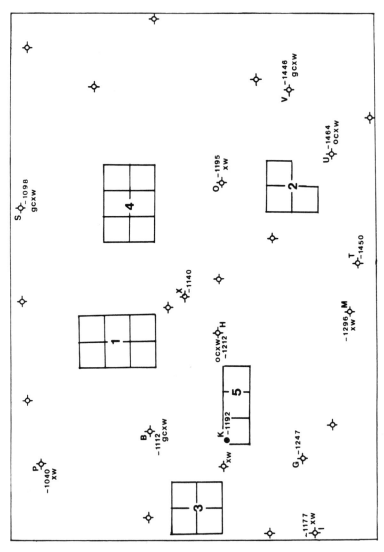

FIGURE 7.33. Base map showing locations of land parcels and wells with pressure data and fluid recoveries.

TABLE 7.2. Data keyed to the locations on the map in
Figure 7.33 for constructing a P–D plot.

Well	Pressure	Elevation
B	2,087	−1,130
G	2,251	−1,260
H	2,227	−1,225
I	2,120	−1,195
K	2,225	−1,205
M	2,269	−1,310
O	2,040	−1,210
P	2,058	−1,060
S	1,988	−1,125
T	2,470	−1,475
U	2,475	−1,485
V	2,465	−1,465
X	2,015	−1,165

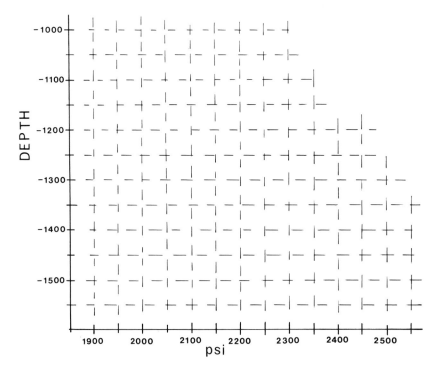

FIGURE 7.34. Graph for constructing P–D plot to map inferred barriers in Figure
7.33.

parallel to the pressure axis, while potential energy (hydraulic head) differences are measured along the gradient normal to the hydrostatic density gradient, as just described. The traces of contour lines on potentiometric maps thus correspond to construction lines that parallel the gradient lines on the graph, and the flow lines that are perpendicular to them correspond to the potentiometric gradient.

Construction of Pressure System Maps

Figure 7.33 is a base map showing well locations, some coded DST results, and several blocks of land. These are from a unit that is suspected of having been sliced by some large regional-scale faults, which create the potential for trapped hydrocarbons. Pressure differences between any fault-bounded systems can also be used to more precisely map them. The pressure data for these wells is presented in Table 7.2, and with this information the reader can construct a P–D gradient plot on Figure 7.34, from which any implied systems can be mapped. On the basis of the results, the land blocks can be evaluated and ranked according to the likelihood of oil pool proximity.

8

Potentiometric Maps and Subsurface Water Flow

The Potentiometric Surface

A potentiometric surface is a calculated imaginary surface, the topography of which reflects geographic variation in the fluid potential of the formation water within a particular aquifer or subsurface reservoir. The elevation of the surface at any point on it reflects (but does not exactly equal) the height to which a column of water would rise above a reference datum within a vertical tube (ignoring capillarity). This is an approximation of the "head," (H_W), as has been explained earlier. The hydraulic head reflects the level of potential energy of the water at the point concerned, and the height of the column mirrors the pressure within the aquifer at that point, as well as the salinity (since pressure is a function of density) of the water. The heavier the fluid, the shorter the column required to balance a particular internal pressure. For example, the height of a column of dense brine required to offset a given pressure in a formation might be only 90% that of a column of fresh water.

In practice, the hydraulic head, H_W, is calculated from pore pressure measurements in the water-saturated rock as follows:

$$H_W = Z + \frac{P}{D_W g} \qquad [8.1]$$

and

$$(H_W - Z) = \frac{P}{grad\ P,} \qquad [8.2]$$

where Z = reference datum in feet above or below an arbitrary but constant datum (usually the subsea-level position of the measurement gauge); P = measured pressure in psi or kPa; D_W = density of the water *throughout the fluid column* above the point of measurement (a tricky parameter to accurately estimate because of the layered nature of geologic rock columns); g = acceleration of gravity; and D_wg = grad P = static pressure gradient.

The relationship between the head and the fluid potential of the water is

$$\phi = gH_w = g\left(Z + \frac{P}{D_w}\right) \qquad [8.3]$$

so that $\phi = gZ + \dfrac{P}{D_w}$ = the potential energy of the water.

For mapping purposes, the hydraulic head H_w serves as a practical approximation of fluid potential ϕ, which differs from it only by the fairly constant quantity g.

Inference of Subsurface Flow Patterns

Figure 8.1 is a schematic cross section showing a rock unit into which four wells have been drilled (A, B, C, and D). The heights of the water columns in the wells are equal, implying that the hydraulic head (fluid potential) throughout the unit is at a constant level. No flow is inferred, the reservoir is hydrostatic, and the trace of the surface represented by the dashed line connecting the four head elevations is horizontal. The formation would be described as normally pressured at the location of well A since the head elevation is close to ground level; overpressured at well B and well D because the head elevation exceeds ground level at both sites; and underpressured at well C since the head elevation is significantly below ground level.

The heights of the water columns in the four wells illustrated in Figure 8.2 increase from left to right, indicating decreasing levels of potential energy in the opposite direction from east to west. And since fluids flow from regions of high fluid potential to those of lower fluid potential, water movement in a westerly direction is inferred.

A local low on the potentiometric surface exists between wells B and C in Figure 8.3, implying that formation water is moving from east to west

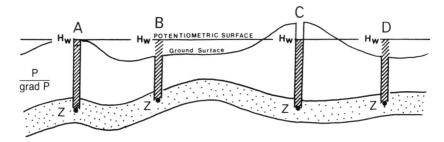

FIGURE 8.1. Cross section illustrating hydraulic head elevations and potentiometric surface based on four wells. Surface is horizontal; thus, no flow is inferred.

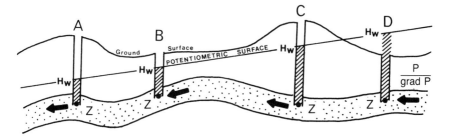

FIGURE 8.2. Cross section showing varying hydraulic head elevations associated with four wells. Potentiometric surface slopes to the west; thus, east-to-west water flow direction is inferred.

in the vicinity of wells C and D and from west to east between wells A and B. Converging flow is inferred from this surface configuration, but for flow to be maintained, the water has to be going somewhere. In such a case, it is most likely moving upward into shallower units, to the surface if conduits such as fractures or faults are available, or even down into deeper formations.

Figure 8.4 illustrates the reverse condition to that just described. A potentiometric "high" between wells B and C implies diverging flow, with water entering the aquifer in the vicinity of the high and spreading out through the unit toward the east and west since the hydraulic head decreases in both these directions. As before, to maintain flow the water has to be coming from somewhere and heading toward some destination (the ultimate potentiometric low?).

The interpretation of local fluid potential differences as reflecting water movement within a single aquifer is valid *only* if there is free fluid transmission throughout the unit concerned. The potentiometric surface in Figure 8.5 implies diverging flow like that in Figure 8.4, but as the cross section reveals, this interpretation is incorrect because, physically,

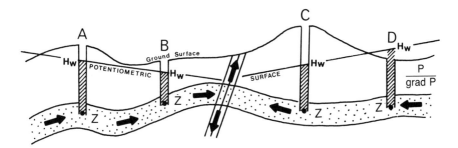

FIGURE 8.3. Cross section reflecting a local low on the potentiometric surface. Converging water flow pattern is inferred.

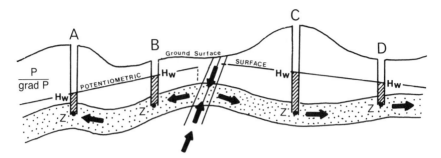

FIGURE 8.4. Cross section illustrating local potentiometric high from which diverging water flow pattern is inferred.

the unit is noncontinuous, with lack of connection between the individual reservoirs. A nonflow scenario would in this case be misinterpreted as representing a flow scenario (water movement). In fact, it is a chance arrangement of individually hydrostatic but physically discontinuous units. (Some of the factors producing varying formation pressures and related potentiometric surface values are described in chapter 6.)

Figure 8.6 is significant because it illustrates an important type of potentiometric feature—namely, the "step." The step reflects the locally anomalous loss of energy over a relatively short flow distance. This results from locally diminished permeability and porosity of the aquifer within which the water is moving, caused in this case by a shaly zone within the sand. The energy loss in the vicinity of this zone of higher resistance to flow is the result of increased friction between the moving water and the reduced pore spaces and throats. The water has to work harder to squeeze through these interstitial spaces among the grains of the rock, and it expends more energy. Usually, such changes in the character of the flow and the corresponding discontinuity on the

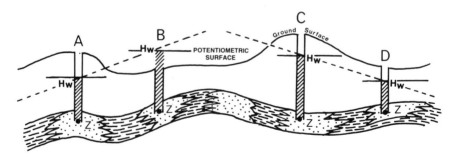

FIGURE 8.5. Cross section demonstrating a local potentiometric high from which diverging water flow pattern might be incorrectly inferred.

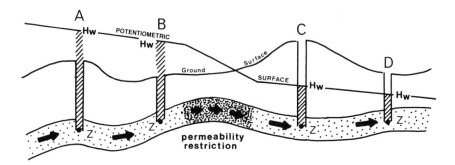

FIGURE 8.6. Cross section illustrating a potentiometric "step" reflecting a water flow constriction due to reduced permeability zone.

potentiometric surface indicate facies changes, boundaries, or fault-related flow restrictions that hamper water movement.

One objective of a hydrodynamic evaluation of a region is to predict locations of hydrocarbon-trapping lithologic barriers by spotting locally radical changes in the general slope of potentiometric surfaces. Therefore, the problems inherent in the mapping of all subsurface data (i.e., density, abundance, and quality of control), contouring procedures, and algorithms are critical in the interpretation, mapping, and display of meaningful potentiometric surfaces. It behooves the maker of the potentiometric map to be extremely careful with machine contouring, because effective anomaly detection requires interpretation. For maximum effectiveness, the map must show a subjective picture of the geologist's interpretation rather than a mechanical representation of the control.

Examples of Potentiometric Maps

Figures 8.7 and 8.8 are examples of two potentiometric maps from Hitchon (1969) based on contoured hydraulic head values from the Devonian Woodbend and Cretaceous (Mannville) groups in the Alberta Basin of western Canada. The contour units are in feet of hydraulic head above sea level, and the pattern implies regional flow from the vicinity of the Rocky Mountain front in the west toward the Canadian Shield to the northeast in both of these units. The potentiometric slopes average about 4–8 ft/mi. A significant feature shown in Figure 8.8 is the prominent potentiometric "sink" showing up in central Alberta. Anomalously, such lows theoretically reflect points where fluids are leaving the system concerned. If the system is dynamic, the low suggests interformational migration with water moving from this unit into underlying Mississippian or Devonian formations or even up toward the surface. Also, production-

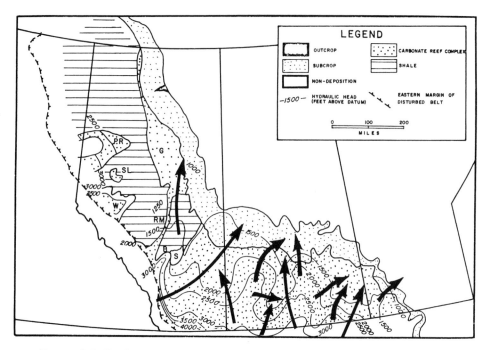

FIGURE 8.7. Regional potentiometric map showing variation in hydraulic head within the Middle Devonian of Alberta. Arrows indicate inferred water flow directions.

related fluid withdrawal over time might explain this intriguing area of low head.

A potentiometric map illustrating geographic variation in hydraulic head (water fluid potential) within the Dakota Sandstone, an aquifer that underlies many of the central U.S. states, is shown in Figure 8.9. The slope of the surface to the east from the region of the Black Hills toward Minnesota and Iowa implies that water is entering the Dakota through the surface outcrop belt that fringes the granitic uplift on the western edge of the map. This map is based on water levels within surface wells.

A regional potentiometric map of Cambro-Ordovician formations from the Algerian Sahara after Chiarelli (1978) is shown in Figure 8.10. Inferred from the figure is that water enters from the surface through the outcrops in the southern part of the area. This water flows north through the units in the general directions indicated by the flow arrows normal to the potentiometric contours. There appears to be a small system to the north reflecting local water movement to the south. Both of these flow systems appear to converge in the center of the area, with deflected water movement off to the west and east, as one might predict. The oil pools in

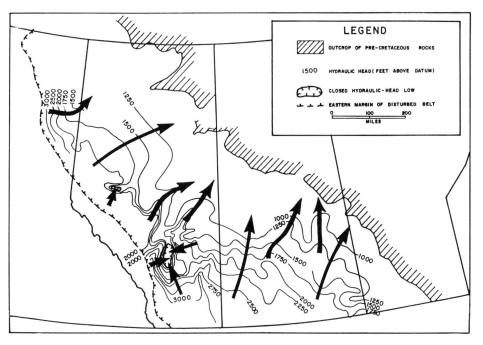

FIGURE 8.8. Regional potentiometric map based on hydraulic head values from the Lower Cretaceous formations. Inferred water flow directions are indicated by the arrows.

the area display nonhorizontal oil–water contacts dipping to the north (see Figures 9.10, 9.11 and 9.12) in the same direction as the regional hydrodynamic gradient.

Figure 8.11, after Dickey and Cox (1977), shows a local-scale potentiometric map based on pressures measured in the Cretaceous Viking Sandstone in Alberta, Canada, the most prominent feature on this map is the potentiometric step reflected in the envelope of contour lines that trends diagonally across the map area, with the hydraulic head averaging about 2,000 ft to the northeast of the step and closer to 1,100 ft to the southwest of the step. The correspondence of the Viking gas pools and the location of the step is obvious. It is not known, though, whether the step exists because of conditions that preceded and thus contributed to the accumulations of hydrocarbons or whether the pools have been trapped by some unrelated mechanism resulting in a damming effect reflected by the step. Nevertheless, exploratory drilling along this trend would have been successful regardless of cause-and-effect ambiguity. If, as well, the step reflects a facies change, it can be assumed that the NE–SW hydrodynamic gradient is probably reinforcing the barrier and in this way enhancing the size of the gas accumulations, so the probability of

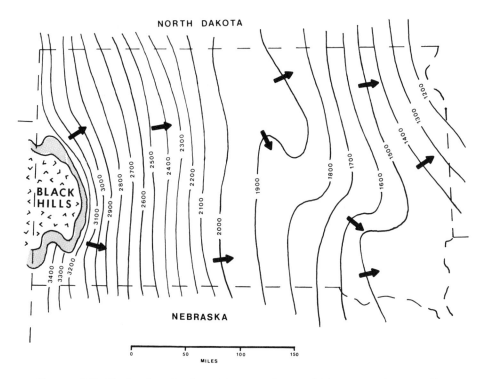

FIGURE 8.9. Regional potentiometric map from which eastward water flow through the Dakota Sandstone aquifer is inferred.

encountering additional gas pools along this trend could be attractively high.

Another example is illustrated in Figure 8.12, which shows the regional potentiometric map for the Gallup Sandstone in the San Juan Basin of New Mexico in the area of the Bisti and Gallegos, oil and gas fields. McNeal (1961) concluded that these oil pools are held by the downdip flow of the fresh formation water combining with the diminished permeability of the seals to impede the updip migration of the hydrocarbons by elevating the oil entry pressure. There appears to be about 3,000 ft of head difference across the map area, with the biggest drop along a NW–SE belt that includes the two fields; a potentiometric step such as this reflects a barrier that undoubtedly facilitates oil entrapment in the two oil pools. Core analysis of the barrier has indicated a *barrier pressure* (also called *entry* or *displacement pressure*) of 22 psi (which is what has to be exceeded for hydrocarbons to move into and through the updip seal). However, the buoyancy force of the oil body creates a pressure that is more than double this amount (about 52 psi). Therefore, the additional 30 psi that makes the seal competent enough to hold back the existing volume of oil is provided by the moving water, as indicated by the flow vectors in the figure. This is hydrodynamic enhancement, since the pools are five times larger than they would be under static conditions! The potentiometric map in this case reveals a facies-stratigraphic trap, because no indication of the pools can be extracted from the structure data. Potentiometric surface slope changes are useful exploration guides in areas where subsurface water flow is apparent, well control is sparse, and facies changes are likely.

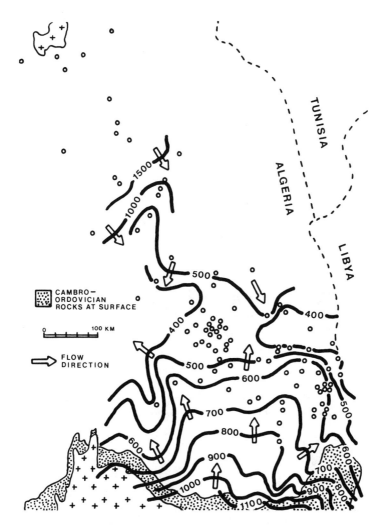

FIGURE 8.10. Regional potentiometric map for the Cambro-Ordovician rocks underlying the Sahara desert in Algeria. Inferred water flow pattern is generally to the north.

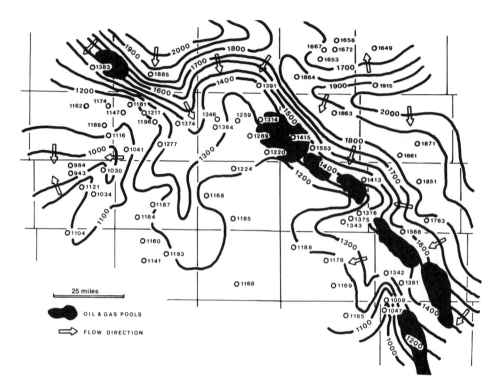

FIGURE 8.11. Local-scale potentiometric map constructed from hydraulic head values in the Cretaceous Viking Sandstone in Alberta, illustrating a pronounced step in the contours.

Regional Potentiometric Maps and Basin Models

The water flow regime in a geologic basin is a function of numerous hydrologic, geologic, and tectonic factors, and the flow patterns can be inferred from regional-scale potentiometric maps. These indicate, in general, broad portions of the basin concerned where conditions appear favorable to entrapment of both oil and gas with respect to size and number of anticipated pools. In other words, hydrologically destructive and constructive areas within a basin may be discerned from regional-scale potentiometric maps.

Figure 8.13 shows in cross section a conceptual model of a basin at a comparatively early time in its history.* The trough has been filled by an influx of sediments, and the deeper units are being pressed upon by those overlying, as illustrated in the figure. Within the single unit shown, the grains are beginning to collapse under the overburden load. The pore fluid pressure is increasing, and in response to the squeeze the fluid

---

* After — & — I.F.P.

migrates in the directions of relief. This is toward the basin margins, creating updip flow conditions (represented by the black arrows). The trace of the potentiometric surface reflects these effects; it is high over the interior of the basin where the hydraulic head is greatest and slopes off at the margins where the hydraulic head within the aquifer is lowest. A schematic P–D plot based on pore pressures at locations A through J in the cross section is shown in the figure. As noted, the deeper measurements are overpressured compared with the shallower ones.

The conceptual hydrology for a more mature basin is illustrated in Figure 8.14. At this stage, the compaction is decreased as the grains within the aquifer of interest begin to take up the weight of the overburden. The pressure within the fluids in the pores relaxes, and fresher water from the surface begins to flow down toward the interior of the basin in response to gravity and the slackening updip flow. Some of the upward-moving formation water can escape up fractures and faults, as shown. The trace of the potentiometric surface reflects these conditions. It slopes down toward the geographic center of the basin and rises again to its highest point over the deeper parts of the basin where the internal pressure is comparatively high. There exist local areas of convergence between the major flow systems, which are reflected as troughs on the

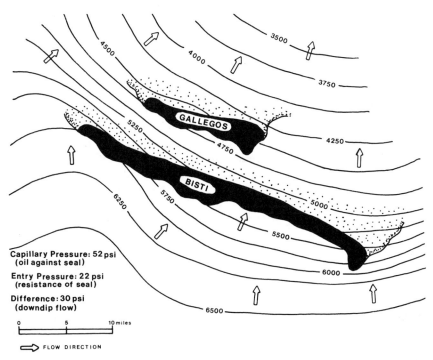

FIGURE 8.12. Potentiometric map in the vicinity of the Bisti and Gallegos oil pools in the San Juan Basin.

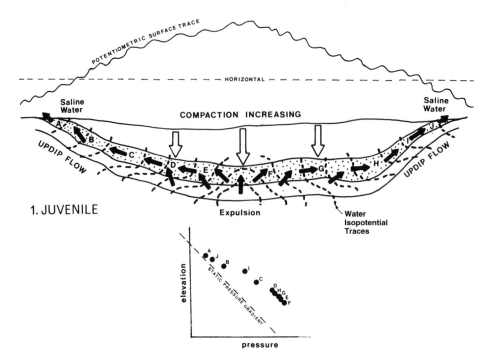

FIGURE 8.13. Conceptualized model of a regional-scale basin in its juvenile stage, showing the relationships among regional flow pattern, potentiometric surface, and the corresponding P–D plot.

regional potentiometric surface. These conditions are likewise reflected in the P–D plot pattern. The pressures measured at the basinal margins reflect the local downdip flow and are also underpressured compared with the pressures deep in the interior of the basin. Remember that on a P–D graph the flow gradient that relates to fluid movement (if there is any) is oriented *normal* to the static pressure gradients, with the flow source up and to the right and the flow destination low and to the left.

At an advanced stage in the life of a basin, compaction for all practical purposes ceases, as illustrated in Figure 8.15. The fluids within the aquifer at the bottom of the basin are stagnant, and if there is any flow it is gravitational from the hydraulically higher margins of the basin toward the interior (i.e., fresh water flowing downdip). The potentiometric surface configuration is a simple one: elevated hydraulic head at the basinal margins grading down into a regional-scale potentiometric low over the deep interior. The flow is downdip, so the marginal slopes of the basin are hydraulically optimal for hydrocarbon entrapment. The corresponding P–D plot shows that the deeper portions of the basin are underpressured with respect to the shallower margins. This hydrodynamic picture pre-

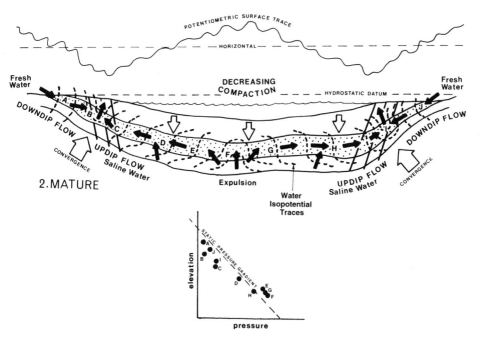

FIGURE 8.14. Conceptualized model of the basin at a more advanced stage, illustrating the relationships among formation water flow, potentiometric surface, and P–D plot.

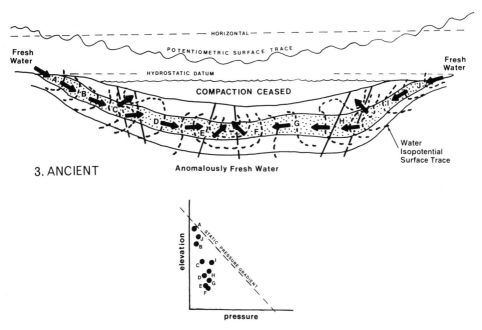

FIGURE 8.15. Conceptualized model of an ancient basin, illustrating relationships among water flow pattern, potentiometric surface, and P–D plot.

sents an enigma (although it characterizes several North American basins) because if the water enters the basin at the margins and flows down into the central depths for flow to be maintained, the fluids have to be going somewhere. In the illustrated case, the water probably migrates into shallower formations along faults and fracture systems.

The fourth model reflects a scenario in which the flow is produced not by compactional processes but by gravity induced flow. As Figure 8.16 illustrates, the basin is regionally tilted and the margin on one side is lower than that on the other, so a directional hydraulic head difference exists. In response, the fluids move unidirectionally through the aquifer from the high side to the low. The regional potentiometric surface tilts in the same direction as the assymetrical basin. The points on the P–D graph reflect this hydrology, with pressure values on the upside occupying the higher positions along the flow gradient (normal to the static pressure gradient) and the pressures measured on the downflow side of the basin in lower positions.

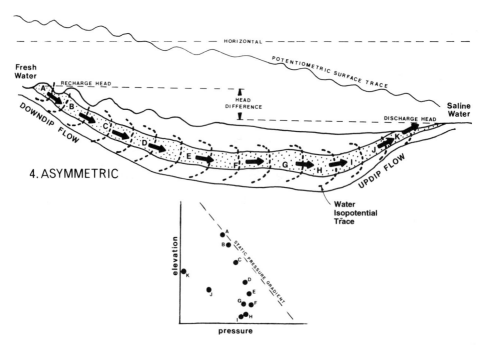

FIGURE 8.16. Conceptualized model of a tectonically tilted basin, showing relationships among water flow pattern, potentiometric surface, and P–D plot with downdip flow on the high side and updip flow on the low side.

Potentiometric Maps of Some Basins

The structure on the base of the Mesa Verde formation in the San Juan Basin of New Mexico is pictured in Figure 8.17. This formation outcrops along the margins of the basin around 5,000 ft above sea level and plunges to below −2,000 ft in the center of the basin, which is about 100 mi across. The Rio Blanco gas field is located in the deepest part. As shown on the

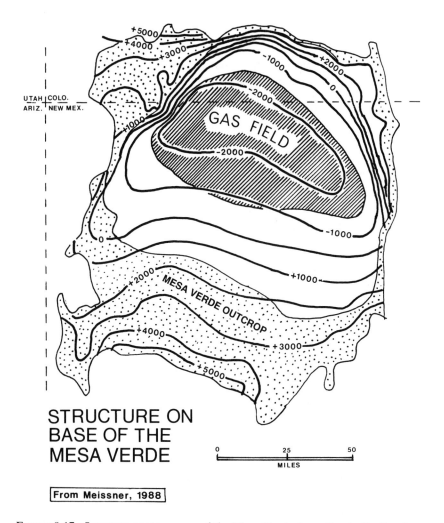

STRUCTURE ON
BASE OF THE
MESA VERDE

0   25   50
MILES

From Meissner, 1988

FIGURE 8.17. Structure contour map of the Mesa Verde formation in the San Juan Basin, showing peripheral outcrop pattern and location of the central Rio Blanco gas field.

map, the outline of the gas closely parallels the basinal structure and is itself over 50 mi across! This bubble sits at the bottom of the basin in apparent defiance of the presently known laws of physics, which dictate that unless constrained by an impermeable barrier the gas should migrate up to basin margins as a function of its natural buoyancy in the water-saturated rocks. However, it doesn't or hasn't, so other factors must be operating to "hold" it in its present position. One of these might be downdip flow, if it exists and if it is intense enough to reinforce any lithologic barriers, thus offsetting the natural buoyancy of the gas.

Figure 8.18 is the potentiometric map for the Mesa Verde formation water; the surface itself mirrors the Mesa Verde structure almost perfectly, with the highs following the outcrop trend along the basin margin and the low coinciding with the deepest part of the basin *and* the Rio Blanco gas field. Implied by this surface is the infiltration of surface water at the outcrop with gravitational flow down into the interior. It has been suggested that the downdip flow might be osmotic—fresh water at the margins separated from highly saline water deep in the center by shaley lithologies acting as semipermeable membranes with a difference in chemical potential across them. The downdip movement of the fresh water, in response to the salinity imbalance, theoretically constitutes the flow inferred from the regional potentiometric map. Available data suggest, however, that the deep-basin brine salinities are too low (compared with those of the shallow and surface waters) to produce a significant osmotic effect of the proper magnitude, so the downdip flow has to be largely gravitational. But if water is entering and moving down with sufficient force to counteract the buoyancy of the gas bubble, where is it going? How is flow maintained without an outlet somewhere? This is a difficult question to answer, although it has been proposed that molecule-by-molecule replacement of the gas by the water (imbibition i.e., absorption of water by gas) might be taking place. When a water molecule "arrives," a gas molecule is released and begins a slow migration upward through the overlying, younger units. In this way, proponents of this idea speculate, flow can be maintained, although over a long period of time the molecular erosion of the hydrocarbons would diminish the overall volume of the gas bubble, unless more gas is replenishing the accumulation from deeper down.

Figure 8.19 is a map of the Madison formation of Mississippian age in the Williston Basin of the western United States. This unit represents the general configuration of this petroliferous basin, which underlies large portions of Montana and North Dakota. Hydrologically, if the basin has been static, any hydrocarbons generated at depth will migrate along buoyancy-dominated paths (normal to the structure contours), as illustrated by the arrows in the figure. The potentiometric map under such conditions would be generally featureless, since the potential energy within the Madison aquifer would be constant.

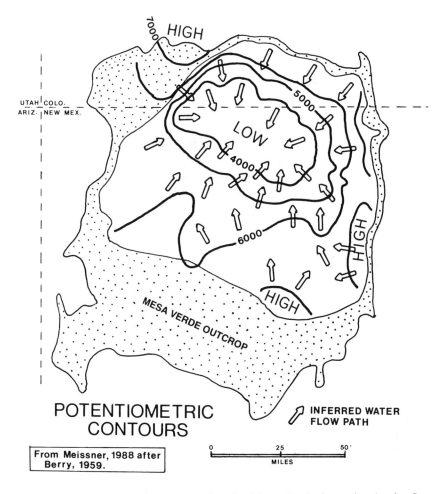

FIGURE 8.18. Potentiometric surface for the Mesa Verde formation in the San Juan Basin, implying downdip flow from the outcrop toward the center of the basin.

The actual potentiometric map for the unit in Figure 8.20 shows that this does not appear to be the case. The potentiometric contours are high to the southwest, decreasing systematically toward the northeast, implying a unidirectional hydrodynamic gradient and flow from the Madison outcrop area up toward North Dakota. The black arrows represent the flow paths that the oil would follow, and the white arrows represent the formation water flow. The dotted zone shows a mass of saline water that has apparently settled to the bottom of the basin as a result of its higher density. The potentiometric contours within this body are based on

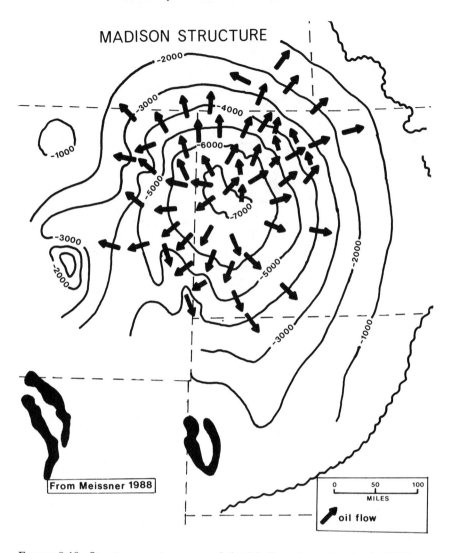

FIGURE 8.19. Structure contour map of the Madison formation in the Williston Basin. The arrows represent flow paths for hydrocarbons in a hydrostatic formation where buoyancy is the only force affecting migration.

hydraulic head values calculated using greater density-gradient approximations in the equation [H = (KB − RD) + P/grad P]. The resulting contour configuration remains relatively unaffected, although the equivalent contours within the "dense" zone are lower in absolute value. (This seems correct since a smaller column of denser water compensates a particular pressure value.)

In a hydrologically static basin, the denser water sinks deeply into the central part of the basin, and its outline will conform to the regional structure contours. When Figures 8.19 and 8.20 are compared, it is evident that the saltwater slug occupies a position that is offset significantly in a northeasterly direction from the structural center of the basin. Hubbert (1953) has addressed the problem of moving fluids of differing densities in contact with each other and has demonstrated that when a

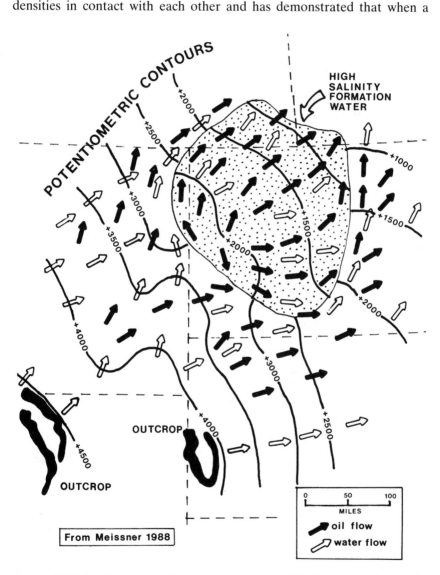

FIGURE 8.20. Regional potentiometric map for the Madison formation in the Williston Basin, implying unidirectional water and hydrocarbon flow and migration from the southwest to northeast.

less dense fluid (such as fresh water) moves over a more dense, relatively static fluid (such as highly saline brine), the contact between the two fluids rotates upward in the direction of flow. This is the opposite of the case in which a more dense water is moving beneath but in contact with the less dense static oil and the interface is tilted downward in the direction of flow. The degree of rotation depends upon the densities of the two fluids and the magnitude of the flow. Therefore, the asymmetrical position of the salty formation water mass reflects the northeasterly flowing water in the Williston Basin.

Figure 8.21 shows a potentiometric map based on pressures measured in the Anadarko Basin in the central United States. The potentiometric values are low along the margins of the basin and increase toward the center or deeper portion of the basin; the potentiometric contours match geographically but are antipathetic to the general structure of the basin. This map can be interpreted as indicating updip flow from the center of the basin out toward the margins, similar to the model of the juvenile basin represented in Figure 8.13.

A potentiometric map of the Salawati Basin in Indonesia, shown in Figure 8.22, suggests regional flow from the formation outcrop areas down into the deeper part of the basin to the west. The extent to which this flow has affected the location and size of associated oil pools and degree of flushing of related geologic features is not known. The potentio-

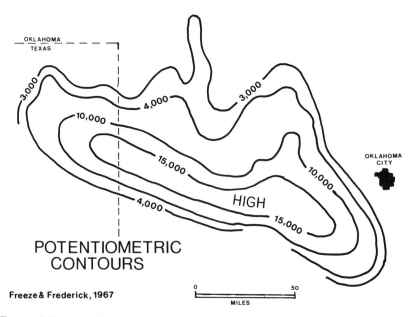

FIGURE 8.21. Potentiometric map for the Anadarko Basin, indicating the probable effects of overpressuring of the units in the deeper portions of the basin.

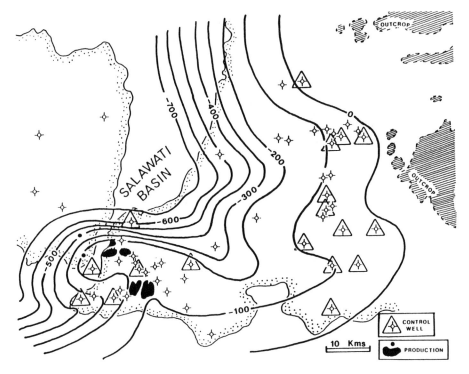

FIGURE 8.22. Potentiometric map from the Salawati Basin of Indonesia, illustrating inferred water flow from outcrop down toward the deeper portions of the basin. (Courtesy of Reid, 1985.)

metric map covering the area of the Indian Basin gas field in New Mexico (Figure 8.23) displays a pronounced low in the vicinity of the field itself; it has the appearance of formation water flow converging toward the gas pool. The critical question is whether the potentiometric low (local anomaly) was present at this location prior to the discovery and development of the pool itself or whether production of gas has created the low.

Local Potentiometric Anomalies and Hydrocarbon Accumulations

For potentiometric maps to be of practical use to petroleum explorationists, there must exist a reliable relationship between map features and the locations of oil and gas pools. If such a relationship exists, then potentiometric maps can be employed for predicting locations of subsurface hydrocarbon accumulations *prior to drilling*. The following model offers a rationale for such an anomaly-target relationship.

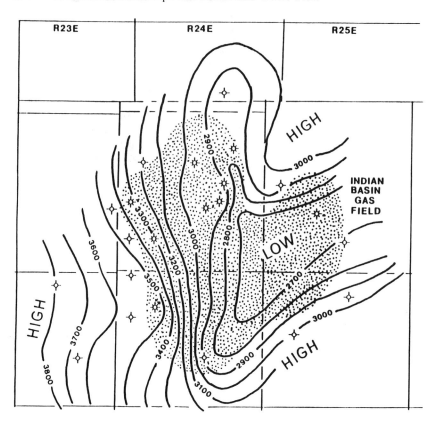

FIGURE 8.23. Potentiometric map showing hydraulic head variation in the vicinity of the Indian Basin gas field, with a pronounced low coinciding with the location of the accumulation. (From Frenzel and Sharp, 1986.)

Figure 8.24 illustrates the theoretical situation of an oil pool located somewhere in a hydrodynamic area; the regional flow pattern is (as indicated by the arrows) northeast to southwest. The heavy black lines are essentially flow lines that reflect the distance that an individual water molecule travels in a constant unit of time. Thus, the length of the line segment is related to flow velocity (i.e., speed). Like a boulder in the bed of a stream, the pool constitutes an obstruction around which the water must move to maintain flow; it can (if aquifer conditions allow) pass over, beneath, or around the pool. No matter which combinations of these paths the water follows, the result will be to create an aberration in the overall flow pattern. If the effect is strong enough, the map control is optimal, and the geologist's interpretive skills are keen enough, the flow distortion shows up on the potentiometric map as a local anomaly. The dashed lines normal to the flow paths represent internal levels of potential energy within the water mass and are analogous to magnetic potentials, which are associated with the lines of force indicated by iron filing patterns on paper over a magnet. These isopotential lines are normal to the flow lines. It is of course impractical and essentially impossible to obtain measurements from drilled wells that reflect local flow velocity, but it is feasible to estimate potential energy level from measured formation pressures. Therefore, the resulting potentiometric map reflects the flow

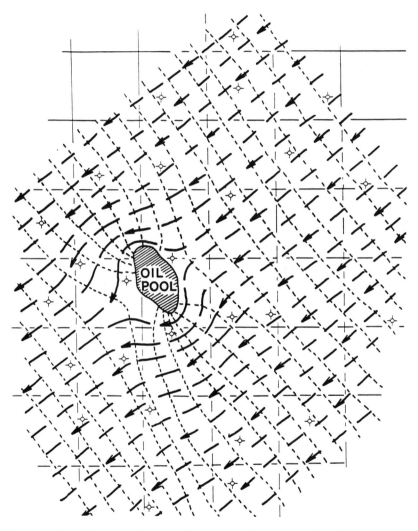

FIGURE 8.24. Schematic diagram illustrating the manner in which a trapped hydrocarbon accumulation should behave like a resistor in a flowing electrical system or a large boulder interfering with moving water in a stream bed. Heavy lines are flow lines and dotted lines represent levels of internal potential energy of the water.

patterns as illustrated in Figure 8.25, which shows the hypothetical potentiometric map generated in this scenario. The anomaly revealing the location of the oil pool is a local "scoop-shaped" low in the potentiometric surface. As with any subsurface map based on widespread control points, success depends on the interpreter's ability to recognize and capitalize on

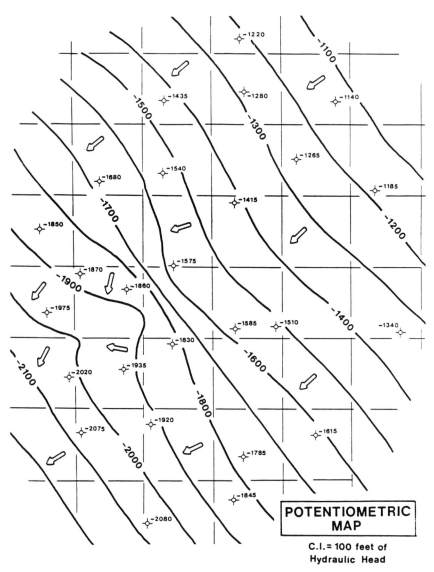

POTENTIOMETRIC MAP

C.I. = 100 feet of Hydraulic Head

FIGURE 8.25. Potentiometric map based on interpretation of hydraulic head elevations calculated from pressures in the indicated wells and resulting "scoop-type" anomaly on the downflow side of the petroleum pool.

the implications offered by the raw data. In producing the final map, it is important to "optimize" the contouring with respect to scoops and local steps on the potentiometric surface. This means one should apply a contouring style that enhances the display of these features at the locations where the control-point clues suggest that they might be present.

Figure 8.26 is a potentiometric map from the Williston Basin area in which there exist a number of producing oil pools. The individual contours representing feet of hydraulic head imply flow toward the east and northeast (from high to low) and the correspondence between the locations of the oil pools and the potentiometric variations. The 3,300 contour is prominently wrapped around the largest accumulation in the central part of the area. The Big Stick, Tree Top, Whiskey Joe, and Franks Creek pools are all located proximal to the large scoop-type anomaly in the potentiometric surface. At the same time, the Beaver Creek pool in the northwestern corner of the map is likewise situated at the head of a scoop-type hydrologic feature. Also, the 3,300 ft potentiometric contour wraps in a convincing manner around the Elk Horn

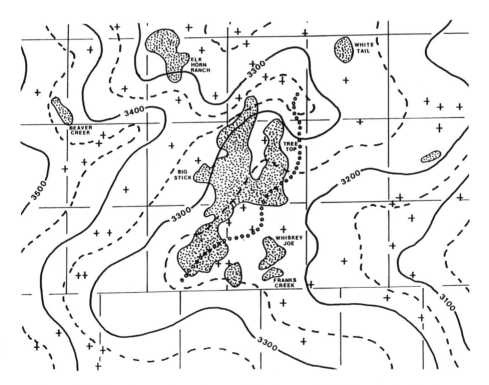

FIGURE 8.26. Local potentiometric map from the central Williston Basin, showing a prominent hydraulic head low (or "scoop-type" anomaly) in the vicinities of the Beaver Creek, Elk Horn Ranch, and Big Stick oil fields. (From Meissner, 1988.)

Ranch pool. Are these patterns coincidental? Or have the potentiometric anomalies been produced by postdiscovery production and fluid removal and therefore are of little use for predictive exploration purposes? It is difficult to tell.

A regional-scale potentiometric map showing variation in hydraulic head within the Muddy formation west of the Black Hills is shown in Figure 8.27. The potentiometric surface drops, in general, from around 4,000 ft above sea level at outcrop to below 1,500 ft near the border. The individual contours are quite irregular, and many of the known oil pools are nestled within scoop anomalies. Some of the anomalies, in which no presently known production exists, are noted by the speckled circles;

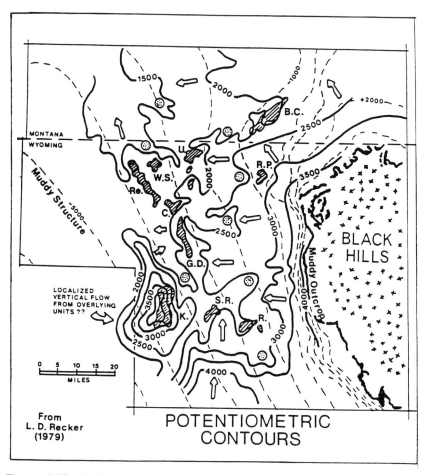

FIGURE 8.27. Regional potentiometric map demonstrating relationship between inferred flow pattern anomalies and the locations of the known, essentially stratigraphic, oil pools in the Muddy formation.

there are seven of these locations. A curious occurrence is the Kitty pool (K.), which lies on a closed potentiometric high rather than a low. Larberg (1981) completed a detailed study of the hydrodynamics of this pool and concluded that the local high reflects incoming water from overlying formations and suggests that this flow pattern has concentrated the oil at this location. A larger-scale view of the Recluse (R.) and West Sandbar (W.S.) pools is offered in Figure 8.28. Note how the contours wrap around the oil accumulation, which sits in the scoop-shaped low. The existence of the West Sandbar pool was purportedly indicated by the associated potentiometric anomaly prior to its discovery.

The potentiometric map for the Bow Island formation in southeastern Alberta, in Figure 8.29, shows variable potentiometric topography with a

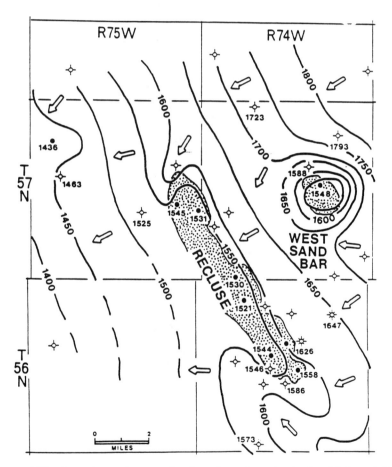

FIGURE 8.28. Potentiometric map showing aberrations in the local inferred flow pattern and the locations of the presently existing Recluse and West Sandbar oil pools.

large step and numerous scoop anomalies, indicating interruptions in the regional flow pattern. Some of these may be attributed to the presence of not-yet-discovered hydrocarbon accumulations. The major anomalies are indicated as the dotted-line circles in Figure 8.30 (for comparison with the locations of the already discovered pools). If other geologic factors are favorable to hydrocarbon entrapment, these locations could be prospective drilling sites with good chances of encountering oil or gas.

A base map in Figure 8.31 is offered to the enthusiastic reader for practice construction of a potentiometric map. There are several dozen

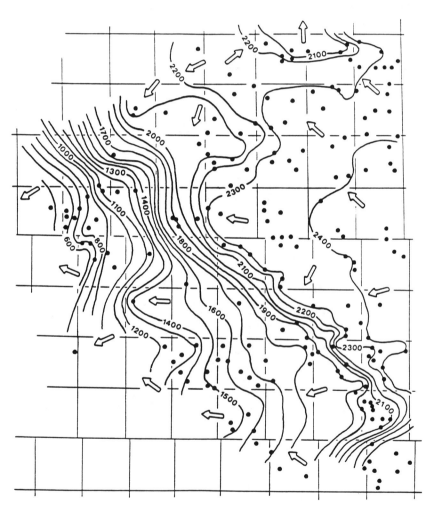

FIGURE 8.29. Potentiometric map from the Lower Cretaceous of southern Alberta showing a prominent potentiometric step and numerous local anomalies. (From Toth and Rakhit, 1988.)

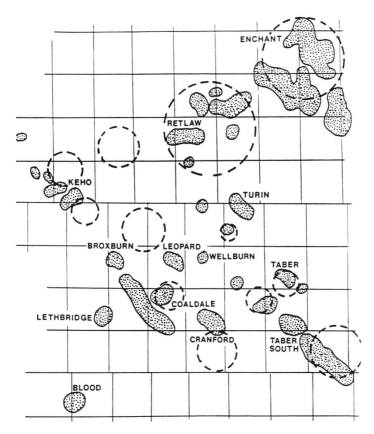

FIGURE 8.30. Locations of existing producing oil pools in the area represented by the potentiometric map in Figure 8.27. Circles represent inferred water flow anomalies.

wells with pressure and depth data in Table 8.1 keyed by letter codes. Remember that contouring is a picture of the geologist's interpretation, so the resulting map should effectively highlight the locations of step or scoop anomalies that are percieved from the data. Hydraulic head values can be calculated using the data in the table and a general gradient value of 0.465 psi/ft or 10.5 kPa/m.

Using a Constant Hydrostatic Gradient Value

In calculating potentiometric elevations from formation pressure data, one must decide the value of grad P to use in the hydraulic head equation. Often there is a water analysis accompanying the reported pressure value, but that reflects only the salinity (and thus density) of the water at the

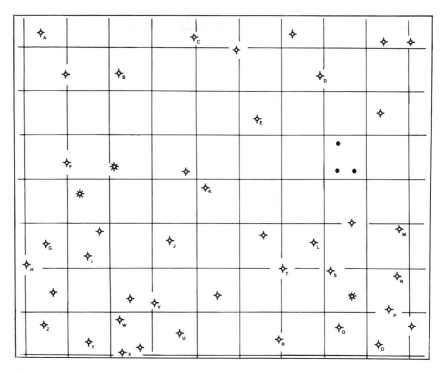

FIGURE 8.31. Base map for construction of a potentiometric map.

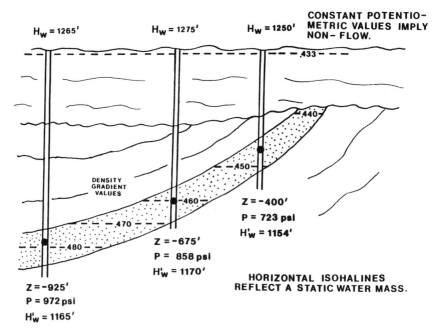

FIGURE 8.32. Relationship between hydraulic head values calculated using true average column gradient values and a general value of 0.465 psi/ft. (Reservoir is hydrostatic.)

TABLE 8.1. Data for the potentiometric map. From this information, hydraulic head values can be calculated and posted on the base map in Figure 8.31 for interpretation and contouring. Hydrostatic gradient is 0.465 psi/ft or 10.5 kPa/m.

Well	KB − RD (ft)	KB − RD (m)	P (psi)	P (kPa)
A	−720	−220	809	5,575
B	−540	−165	762	5,251
C	−185	−56	691	4,762
D	−255	−78	792	5,458
E	−325	−99	745	5,134
F	−475	−145	650	4,479
G	−680	−207	644	4,438
H	−755	−230	648	4,466
I	−670	−204	602	4,149
J	−630	−192	735	5,065
K	−345	−105	676	4,659
L	−350	−107	717	4,941
M	−155	−47	693	4,776
N	−50	−15	592	4,080
O	−165	−50	541	3,928
P	−115	−35	521	3,590
Q	−250	−76	519	3,577
R	−260	−79	500	3,446
S	−245	−74	554	3,918
T	−390	−119	667	4,597
U	−570	−174	643	4,431
V	−550	−168	637	4,390
W	−670	−204	587	4,045
X	−740	−226	631	4,349
Y	−755	−230	601	4,142
Z	−925	−282	656	4,521

point of measurement, not the average gradient over the entire water column overlying the point of interest. It is the latter that is required in the calculation, however.

Therefore, we recommend that one utilize two general values: 0.465 psi/ft if it is a briney area and 0.435 psi/ft if the regional water is likely to be significantly fresh. The possible effects of doing this are illustrated in the following figures. Figure 8.32 shows a static aquifer unit with the salinity decreasing up through the unit toward the surface. The H_W values express the "true" hydraulic head elevations calculated from the weighted average gradient for the whole water column. They are all quite close to 1,250 ft, and from these three values one would infer a horizontal potentiometric surface (thus a static reservoir). The H_W' values are the calculated hydraulic head values using a 0.465 general static gradient value with each pressure rather than the column-averaged (true) values. They all lie close to 1,160 ft, and from them one would likewise conclude

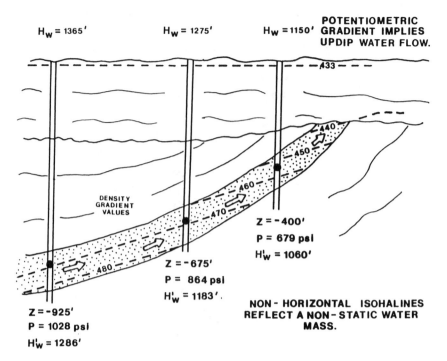

FIGURE 8.33. Relationship between hydraulic head values calculated from the actual water column values and the general 0.465 psi/ft grad P parameter. (Aquifer is hydrodynamic.)

that the unit was hydrostatic. Thus, only the absolute value is affected by the use of a general gradient value.

The case of a *nonhydrostatic* reservoir is illustrated in Figure 8.33. The true water-column-gradient values indicate an almost 200 ft head drop across the three wells, and thus a tilted potentiometric surface. From this one would infer a dynamic aquifer with internal water flowing from west to east. The same conclusion would be drawn from the hydraulic head values calculated using a general gradient value of 0.465 psi/ft, as illustrated. The head drop to the east would still be about 200 ft, so although the absolute head values are affected the relationship between the individual values remains correct. In real life, the overall gradient value up through thousands of feet of rock is impossible to determine, so things are easier when the general values are utilized (unless there is a good reason not to, which rarely is the case).

9

Tilted and Displaced Oil
and Gas Pools

Nonhorizontal Hydrocarbon–Water Contacts

Hubbert (1953) has demonstrated that one of the effects of moving water upon a trapped accumulation of oil or gas in a reservoir is that the interface between the water and the oil tilts downward in the direction of flow, creating an inclined hydrocarbon–water contact, or transition zone. Furthermore, he showed that the slope on the potentiometric surface in the vicinity of the hydrocarbon accumulation that reflects the water flow is numerically related to the tilt on the hydrocarbon–water contact by a factor determined by the contrast between the densities of the underlying dynamic water phase and the overlying static oil or gas phase.

The parameters are illustrated in Figure 9.1, which shows two wells a distance X units apart, both of which penetrate the same formation and pass through the same hydrocarbon–water contact, which is at a shallower depth in the updip well than in the downdip well. The oil–water contact is inclined, and the tilt is the difference in elevation over the distance between the wells, which is $\Delta Z/\Delta X$. The hydraulic head, since the flow is from east to west, is greater in the updip well (2) than in the downdip well (1). The slope on the potentiometric surface in the vicinity of the two wells is thus the difference in hydraulic head elevation over the distance between the wells, or $\Delta H_W/\Delta X$.

The tilt and the slope are related by a quantity called the *tilt amplification factor* (TAF), which is equal to the ratio of the water density to the difference between the water and hydrocarbon densities. Thus, to estimate the tilt of a particular hydrocarbon–water contact using the slope of the potentiometric surface in the locality concerned, the following expression is utilized:

$$\begin{matrix} \text{Tilt of} & & \text{Tilt} & & \text{Slope of} & \\ \text{hydrocarbon–water} & = & \text{amplification} & \times & \text{potentiometric} & \quad [9.1] \\ \text{contact} & & \text{factor} & & \text{surface} & \end{matrix}$$

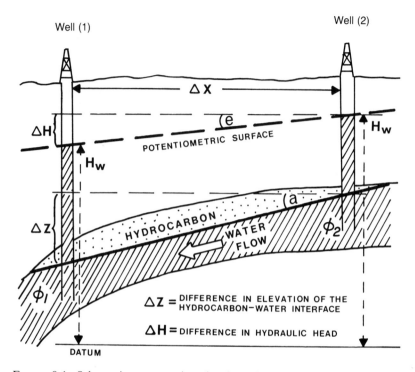

FIGURE 9.1. Schematic cross section showing relationship between magnitude of slope of the potentiometric surface (angle e) and the degree of tilt of the related hydrocarbon–water interface (angle a).

$$\frac{\Delta Z}{\Delta X} = \frac{D_W}{D_W - D_H} \times \frac{\Delta H_W}{\Delta X}, \qquad [9.2]$$

where D_W = water density and D_H = hydrocarbon density (oil or gas).

Table 9.1 lists tilt amplification factors for some hydrocarbons with varying gravities in contact with waters of varying salinities.

The hydrocarbon–water interface is a plane along which the level of potential energy is constant, so it parallels internal isopotential planes within the formation water-saturated portion of the reservoir. The condition required to create tilting is that the more dense water be in motion beneath and in contact with the less dense hydrocarbon phase, which is held in place by the flow, buoyancy force, and the walls of the trap.

As an example, for an oil with a density of 0.87 gm/cc and flowing water of density 1.05 gm/cc, the TAF would be 5.8, which means that the oil–water contact would physically dip in the direction of flow six times more steeply than the slope of the local potentiometric surface. If the

TABLE 9.1. Tilt amplification factors for varying combinations of hydrocarbon density and brine salinity.

API oil gravity		Water salinities		
		200,000 ppm	100,000 ppm	15,000 ppm
		71.7 lb/ft³	66.8 lb/ft³	63.05 lb/ft³
20°	58.3 lb/ft³	5.5	7.9	13.3
30°	54.7 lb/ft³	4.3	5.5	7.5
40°	51.5 lb/ft³	3.6	4.4	5.5
50°	48.7 lb/ft³	3.2	3.7	4.4

overlying hydrocarbon phase were gas with a density relative to water of 0.4 gm/cc, the gas–water interface would be 1.6 times steeper than the corresponding potentiometric surface slope. The significance of a hydro-dynamically-induced tilted hydrocarbon–water contact on oil and gas entrapment is demonstrated in the following figures.

Figure 9.2 is essentially a cross section through a nonclosed monoclinal structure that under static conditions could not hold hydrocarbons. The trace of the regional potentiometric surface illustrated at the top of the figure slopes downward to the east at about 150 ft/mi. If the densities of the local water and hydrocarbon phases are such that the TAF is near 2, there could not be a closure between any hydrocarbon–water contact and the top of the reservoir unit, so any oil would migrate updip from the nose. With a TAF of 3, a very small accumulation could be held in the top of the structure, as indicated by the cross-hatched area in the diagram. With a TAF approaching five, a significant accumulation can be trapped in the structure since the tilted oil–water contact closes against the reservoir seal at the base of the nose and at the top of the structure. If enough hydrocarbon migrates into the structure while these conditions dominate, the size of the largest possible pool is proportional to the area of the cross-hatched portion in the diagram.

With a TAF of 8, the attitude of the hydrocarbon–water contact is such that because of the increased dip, the intersection of the contact and the top of the reservoir at the updip end moves toward the nose as the plane rotates around the downdip spill point. This reduces the size of the largest possible accumulation and represents incipient flushing, which eliminates more and more of the oil accumulation as the hydrocarbon–water contact tilts farther and farther. When the dip on the fluid contact finally exceeds that on the front of the structure, the reservoir is completely flushed and can hold no oil or gas at all. Any movable hydrocarbon phase under these conditions would be carried out of the structure by the formation water flow. As demonstrated, a hydrodynamic component can be constructive up to a point, beyond which it becomes destructive to entrapment, depending on the intensity of the water flow (not necessarily velocity!) as

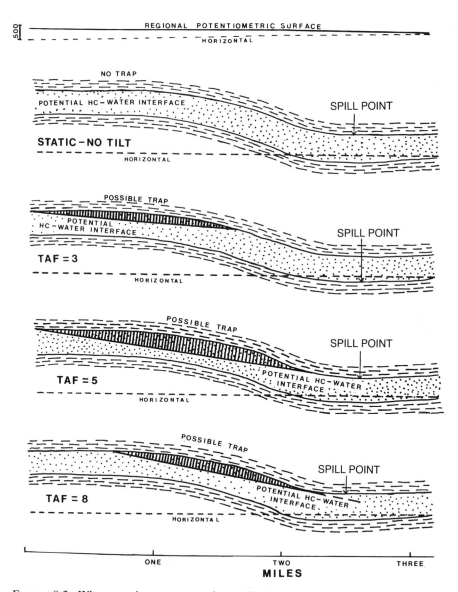

FIGURE 9.2. When moving water produces tilted hydrocarbon–water contacts, nonclosed geologic features such as this monocline become potential accumulation sites. The relationship between volume of trapped hydrocarbon and TAF under constructive and destructive hydrodynamic conditions is illustrated.

reflected by the potentiometric gradient, the appropriate TAF, and the prominence, geologic completeness, and competence of the feature itself.

A tilted oil pool is illustrated in Figure 9.3. This is the Norman Wells field in the Northwest Territories, the limits of which far exceed the volume of the closure shown on the structure. When a rough field outline is sketched in and the points of equal oil–water contact elevation are connected with contour lines, the result presents a map of the contact, as shown in the figure. Adjustment of the contours (which represent a roughly planar surface dipping to the south) by refining the shape of the pool discloses locations for step-out wells. The tilt on the contact appearing on the map suggests a dip of about 400 ft/mi. If the density-dependent TAF in this case were 5, one might reason that the potentiometric surface slope in this area is close to 80 ft/mi to the south. It is not known for certain whether the Norman Wells pool is actually affected by hydrodynamic flow. Some authors have suggested that it is (Hubbert, 1953); others do not agree.

The relationship between API gravity of the hydrocarbon phase and resulting contact tilt is illustrated schematically in Figure 9.4, which shows variation in orientation of principal force vectors for hydrocarbons with API gravities of 20°, 40°, and 80° and the corresponding hydrocarbon–water interface orientation for each. (Remember that the isopotential energy planes are oriented normal to the force vector in this construction, as described in chapter 3.) The greater the API gravity, the closer to

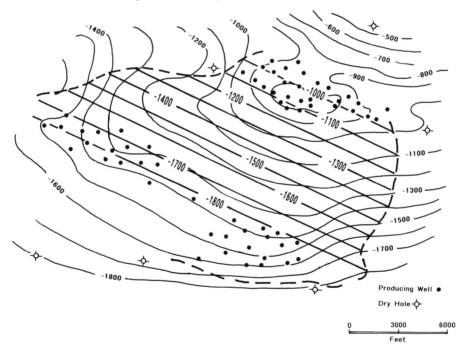

FIGURE 9.3. The Norman Wells oil pool in the Northwest Territories. When the pool boundaries are outlined, a tilted oil–water contact can be contoured, which suggests a local hydrodynamic flow to the south with a potentiometric slope of about 80 ft/mi.

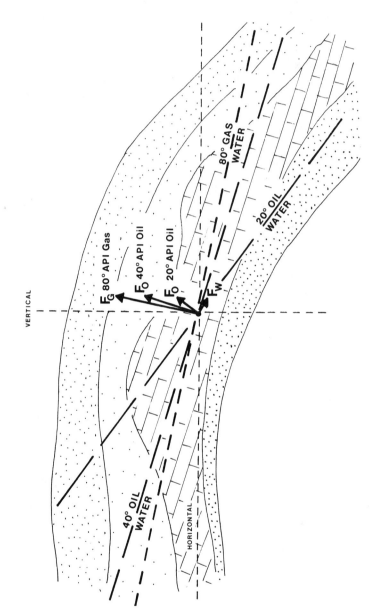

FIGURE 9.4. Magnitude of tilt of hydrocarbon–water contact increases as API gravity of the hydrocarbon phase decreases (the hydrocarbon becomes heavier). Vectors represent the net force acting on the hydrocarbon phase.

horizontal in a hydrodynamic reservoir is the contact. For the carbonate feature in the figure, at a point between 40° and 20°, the tilt would reach the flush point, so medium to heavy gravity crudes could not be trapped in this feature under these conditions.

Tilted Oil–Water Contacts Attributable to Factors Other Than Hydrodynamic

Other mechanisms besides moving formation water can produce variation in hydrocarbon–water contact elevation. These include originally horizontal contacts in a formation that has been tilted by more recent tectonic events, at a rate more rapid than that to which the oil–water contact is able to adjust, and variation in capillary pressure as a function of sediment pore size.

When three glass tubes of different diameter are suspended in a vessel half full of water and half full of oil in such a way that the tubes cross the oil–water contact (as shown in Figure 9.5), discrepancies in the height of the hydrocarbon–water interfaces within the tubes can be observed. Since

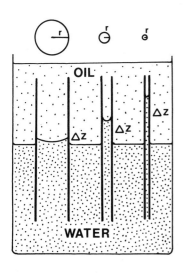
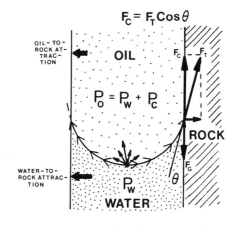

$$F_C = F_T \cos \theta$$

$$P_O = P_W + P_C$$

ΔZ = Height of Capillary Rise

P_C = Capillary Pressure Component

$P_{O,W}$ = Pressure in Oil and Water

F_T = Surface Tensional Force

F_C = Capillary Force Component

F_G = Force of Gravity

ΔZ is proportional to $1/r$, F_T, $\cos \theta$, P_C

FIGURE 9.5. Relationship between height of rise of water against oil (ΔZ) in tubes of decreasing diameter, illustrating the effects of capillary force (F_C) on the position of oil–water and gas–water contacts.

the two phases are immiscible, the interfacial tension between them is high. (This is the tendency for each type of molecule to be more strongly attracted to those of the same kind on the one side of the interface than those of the opposing kind across the interface.) The angle between the wall of the tube and the interface is essentially a function of this as well as the intermolecular attraction between the oil and water molecules and the wall itself. Generally, rocks are water wet, which means that the water molecules are more strongly pulled against the mineral pore walls than the oil molecules and thus will displace them along the wall. Because of the contrast between the water–mineral and the oil–mineral attraction, there is a strong tendency for the spherical interface to try to flatten itself out (thus minimizing the surface area of the fluid interface) by tensional pulling. The resulting force vector F_T is resolved in a direction parallel to the mineral wall F_C, as shown in the figure. The water is pulled up the tube as long as F_C can pull the water column against gravity (F_G). Since pressure equals force per unit of area, the smaller the area of the interface, the greater the pressure difference across the fluid interface between the water and the oil. This difference is the capillary pressure (P_C). The pressure in the water is (P_W), and the pressure in the oil is (P_O), so that at equilibrium the position where ΔZ is a maximum

$$P_W = P_O + P_C. \qquad [9.3]$$

Thus, the position of the oil–water contact in the large-diameter tube does not differ greatly from that outside the tube, while in the small-diameter tube the difference (ΔZ) is extreme. Since a sedimentary rock can be thought of as a network of an almost infinite number of intergranular tubes, the overall position of an oil–water contact will be a statistical average of the sum total pore throat size for the rock concerned. Average pore throat size is proportional to overall grain size (smallest in silts and clays and largest in gravels and conglomerates).

Figure 9.6 illustrates an essentially nonhorizontal hydrocarbon–water contact related to a systematic grain-size gradient. The corresponding differences in capillary pressure vary from a high of up to 40 atm in a clay-size sediment to a low approaching 0.002 atm in granule-size material. The result is a tilted contact, as shown in the figure. Calculations indicate that the maximum amount of vertical displacement producible by pressure differences over this pressure range is on the order of 50–60 ft. If such a discrepancy was observed in two wells a quarter of a mile apart, the dip of the oil–water contact would be interpreted as about 250 ft/mi and thus hydrodynamic (erroneously). Actual oil–water contacts that dip on the order of 600–800 ft/mi are observed in known hydrodynamic basins, so while capillarity accounts for some of the non-horizontality, moving water appears to be the major cause.

The close relationship between water flow pattern and direction of oil field tilting is effectively demonstrated in Figure 9.7, which is a potentio-

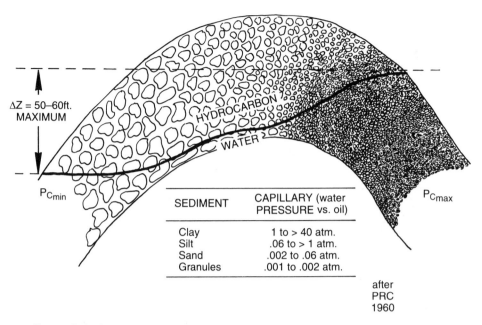

SEDIMENT	CAPILLARY (water PRESSURE vs. oil)
Clay	1 to > 40 atm.
Silt	.06 to > 1 atm.
Sand	.002 to .06 atm.
Granules	.001 to .002 atm.

after
PRC
1960

FIGURE 9.6. Cross section showing tilted oil–water contact in a sedimentary unit produced by systematic change in grain size (pore size). Water exerts greater force on the overlying hydrocarbon phase, where the grains and intergranular spaces are the smallest.

metric map based on hydraulic head data from the Tensleep Sandstone in the Big Horn Basin of Wyoming. The water flow patterns within the Tensleep are inferred from the solid contours, which suggest recharging at the outcrop (potentiometric elevation equal to outcrop elevation), with the water moving into the basin in directions indicated by the white arrows. Of significance is the way that the oil–water contacts in the ten pools illustrated tilt in the direction of implied water movement. From this evidence it is difficult to deny that a cause-and-effect relationship is apparent. Tilted oil pools in the Big Horn Basin include the Murphy Dome pool, with an O-W-C tilt of approximately 200 ft/mi and potentiometric surface slope of 35 ft/mi; the Grass Creek field, with a tilt of 700 ft/mi; Lake Creek, with a tilt of 250 ft/mi; Sage Creek, with a tilt approaching 800 ft/mi; and Frannie, with a tilt of 500–600 ft/mi.

Another documented tilted oil–water contact is displayed by the 200 million-barrel Joarcam pool in Alberta, shown in Figure 9.8. At the northern end of the pool the oil–water contact elevation is close to −727 ft. In the middle of the pool the contact has dropped to −777 ft and at the southern end the same contact is around −815 ft. Thus, there is a nearly 100-ft discrepancy from one end of the pool to the other. Whether this tilt can be attributed uniquely to hydrodynamics or to tectonic tilting,

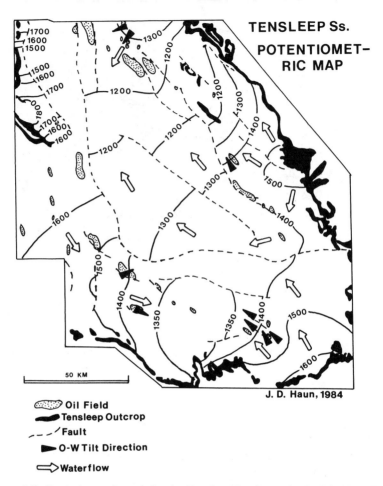

FIGURE 9.7. Potentiometric map for the Tensleep Sandstone in the Big Horn Basin in Wyoming, demonstrating the correlation between potentiometric gradients and related tilted oil–water contacts in associated pools.

as has been proposed by Edie (1955), is difficult to determine, although the subsurface water flow pattern inferred from the local potentiometric map in the previous chapter (Figure 8.11) is compatible with the direction of tilt, as theory proposes that it should be.

The North Tioga oil field is shown in both map and cross section in Figure 9.9, with a contoured dipping oil–water contact. An elevation difference of 400 ft is apparent. As the structure contours show, this nonclosed formation would be unable to hold oil or gas under static conditions.

Figures 9.10 and 9.11 show maps of three oil pools in the Algerian Sahara that are associated with minimally closed reservoirs. The mapped oil–water contacts dip toward the north, and potentiometric maps of the region suggest that the regional formation water flow in these units in the same direction as demonstrated on the potentiometric map in Figure 8.10 is likewise to the north, so the attitude of the fluid interfaces reflect the flow, which undoubtedly has created these nonhorizontal contacts. These pools are hydrodynamic traps that could not exist without the regional flow gradient to the north.

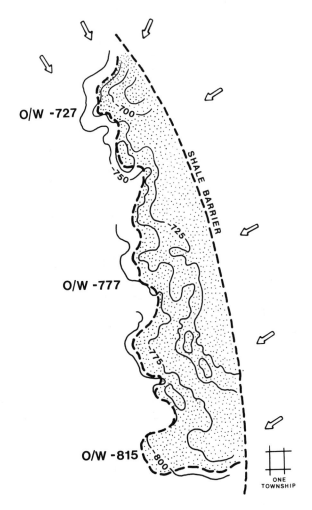

FIGURE 9.8. Map of the Joarcam Viking pool in central Alberta, illustrating an oil–water contact that occurs at −727 ft at the north end of the pool, dropping to −815 ft at the south end. Local potentiometric gradients are represented by the arrows.

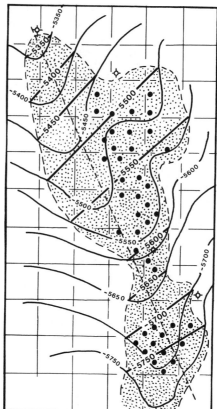

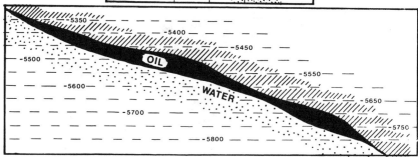

FIGURE 9.9. North Tioga oil field, showing contours on the southward-dipping oil–water contact. Accumulation is held in a nonclosed nose, presumably by water flow in the Mission Canyon formation from north to south.

Economic Significance of Tilted Oil–Water Contacts

Hydrodynamically induced tilting not only allows accumulations of hydrocarbons to exist in geologic features that lack the closure to hold them under static conditions but also has a strong effect on trap capacity. The anticlinal reservoir in Figure 9.12 is 100% full when the oil–water contact is coincidental with the spill point to the right. Any additional oil would be displaced around the spill point and would migrate updip. The reservoir is hydrostatic, and the oil–water or gas–water contacts are horizontal. If the formation is not hydrostatic (therefore hydrodynamic) with water flow in the updip direction to the right, any oil–water contact will be

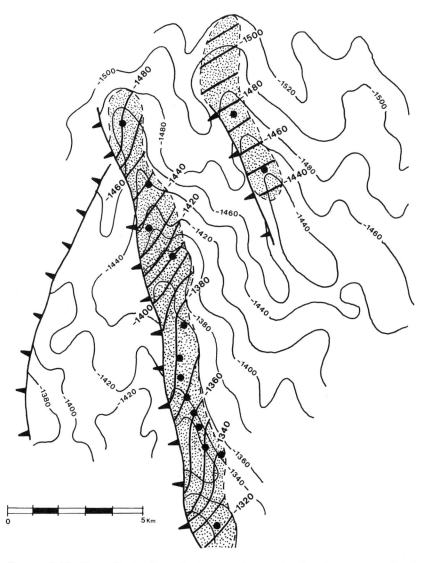

FIGURE 9.10. Two tilted oil pools trapped in poorly closed structures in the Devonian of the Tin-Fouye-Tabankort region of the Algerian Sahara. Contours on the oil–water contacts demonstrate dip to the north coincidental with regional flow reflected on potentiometric maps.

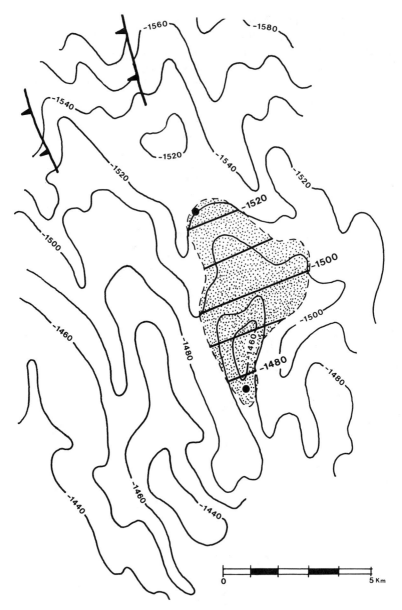

FIGURE 9.11. Another northwardly tilted oil pool in the Devonian of the Algerian Sahara. Regional flow is likewise to the north.

tilted in the direction of water movement (in this case, updip), resulting in the diminishing of the volume of any accumulation as the oil–water contact rotates around the upside spill point. Any additional oil finding its way to this location would move around the spill point and on updip. If the flow is downdip, however, the oil–water contact rotates around the spill point in the downflow direction, with a resulting increase in the maximum volume of hydrocarbons that the trap can hold, as illustrated. Thus, downdip hydrodynamic flow with downdip tilted oil–water contacts

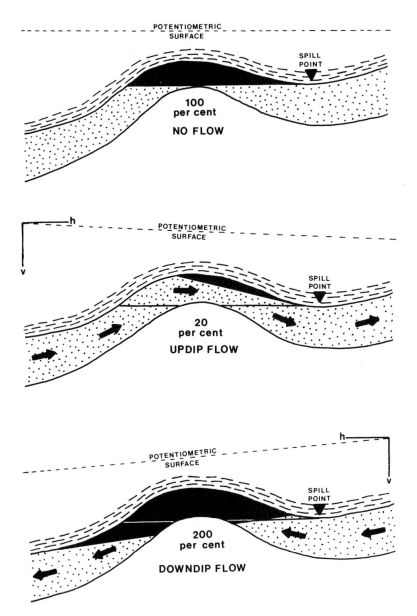

FIGURE 9.12. Variation in trap capacity created by oil–water contact tilt as a function of formation water flow direction.

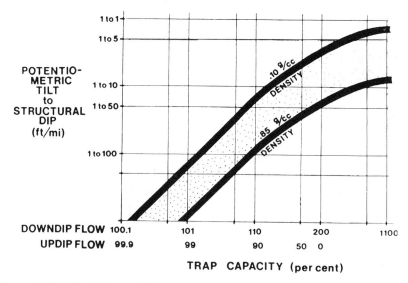

FIGURE 9.13. Graph for predicting trap capacity in terms of potentiometric surface slope (flow intensity), structural dip of the reservoir, and density of the hydrocarbons for updip and downdip flow conditions. (After Schowalter from Berg.)

is favorable and accumulations can be expected to be greater in size than in updip flow situations, which diminish rather than enhance trap capacity.

In a heroic attempt to quantify the interaction among these factors (trap capacity, water flow intensity, and densities of the oil, gas, and water), Schowalter (1979) produced the graph in Figure 9.13. This graph expresses the relationship between potentiometric surface slope, structural dip, and trap capacity as a function of hydrocarbon density for updip (destructive) and downdip (constructive) flow conditions. For example, if the slope of the potentiometric surface is one-tenth that of the structural dip on the reservoir, and the density of the oil concerned is around 0.85 gm/cc (if the direction of flow is updip), the structure will be unable to trap the hydrocarbons. A capacity less than zero is read on the horizontal axis. At the same time, if the flow is downdip, under these circumstances the accumulation would be two or three times larger than under static conditions (200–300%).

U, V, Z Construction Procedure

The principles described by Hubbert and the equations developed by him can be utilized for more specific analysis of the ability of geologic features to hold hydrocarbons under varying sets of geologic and hydrologic conditions. This is accomplished by calculating what the potential energy level of the oil or gas would be at specific points in the subsurface if it were actually present, as a function of the hydrogeologic parameters.

It is not possible to characterize the orientation within a pore-space-water-saturated reservoir rock directly or by observation of the force vectors that control the behavior of the oil, water, and gas phases. However, it is possible to determine the orientation of the isopotential energy planes, and as demonstrated in chapter 1, the vectors are oriented normal to these planes. These forces determine the three-dimensional trajectories and resulting migration paths of the water, oil, and gas and can be illustrated in cross sections and on maps.

The initial requirement is to express the oil potential ϕ_O (which cannot be measured directly) in terms of ϕ_W, which can be determined from the potentiometric map which is based on hydraulic head values (H_W) calculated from formation pressure measurements.

The family of surfaces defined by and passing through similar values of ϕ_O which is the potential energy level for oil (actually H_O), are referred to as oil isopotential surfaces (U). V surfaces calculated from the H_W values are water isopotential surfaces required for calculation of the U values. The orientation of the resulting U surfaces then reflect how oil or gas will behave under the particular hydrogeological conditions concerned. The Z values are actually reference points of constant elevation above or below a specified datum. These correspond to contour slices, with which exploration geologists are familiar.

Mapping Oil and Gas Potential Energy Levels from Water Pressures

The complete "applied hydrodynamics" algorithm is displayed in Figure 9.14. The first step of the procedure utilizes formation pressure P, and with this parameter P–D plots are constructed that predict the positions of oil–water contacts, show correlations between units (thus exposing reservoir trends), and indicate the presence of associated permeability barriers. From P, using the formula for H_W, the hydraulic head of the water within which the pressure has been measured is calculated. From these values distributed over an area of interest, the potentiometric map is constructed, and from it subsurface water flow patterns are inferred, anomalies are identified, and facies boundaries are recognized. At this point in the sequence, the densities of the hydrocarbon phases enter the picture.

If the hydrocarbon phase is oil, the parameter V_O, which is referred to as the water isopotential with respect to oil, is calculated by multiplying the density-related TAF term by the hydraulic head elevation, H_W.

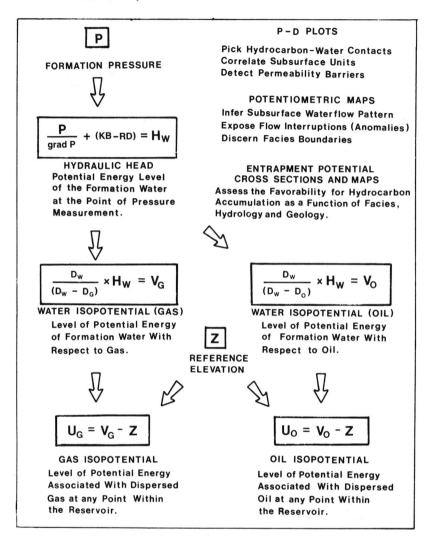

FIGURE 9.14. Flow diagram illustrating the complete hydrodynamic evaluation algorithm, commencing with formation pressure measurements and ending with entrapment potential constructions. Effects of hydrology, facies, and structure are included in these parameters.

Numerically, this calculation yields an elevation on a hypothetical oil–water contact, within the particular unit involved, but the principal value of this parameter is that it is a calculational step leading to the ultimate oil or gas isopotential parameters, U_O or U_G, which cannot be measured directly as H_O or H_G. As illustrated in the figure, U_O reflects the level of potential energy that would be associated with any oil at a particular point within the water-saturated reservoir rock. (Fred Meissner has pointed out that when Z is an elevation on the roof of the reservoir, $V_O - Z$ becomes the isopach of the "possible pay," the interval from the top down to the tilted oil–water contact.) The internal orientation of the planes of constant potential energy for oil or gas can be interpreted from the three-dimensional grid of these values within the formation concerned. In cross sections, the U lines are essentially the traces of the planes of constant potential energy for oil as described above, and on maps the U contour lines are traces of the U surfaces as viewed from above. Entrapment potential maps and cross sections from which the migration paths followed by the hydrocarbons can be interpreted are based on these. The most likely locations of accumulations in terms of hydrology (which includes the effect of facies) geology and densities of the hydrocarbons and the ubiquitous formation water through which the forces are transmitted can be predicted from these as discussed in the next chapters.

10

Entrapment Potential Cross Sections

Hydrostatic Structural Trap

Figure 10.1 shows a cross section through a simple anticlinal reservoir. The Z values represent relative elevations above a fixed datum. These are reference elevations at which levels of potential energy for oil or gas can be easily calculated, and the points at which the Z traces intersect the top of the unit correspond to structure contours. The "contour interval" is ΔZ. Since the system is hydrostatic, the buoyancy vector and the gravity vector representing the forces acting on a unit mass of water offset each other, and the potential energy level of the water is constant. The lowest pore pressure, within the formation water, is at the highest point in the reservoir, which is the crest of the structure. The equipressure planes are oriented horizontally, as illustrated by the dashed lines in Figure 10.2.

However, for oil, there *is* a force imbalance since the upward force component ($-\text{grad } P/D_O$) exceeds g, the force of gravity. The white arrow represents the buoyancy force acting on a unit mass of oil in this static environment. Since buoyancy is the major force and it operates vertically (as does gravity), the planes of constant potential for any oil under these conditions are oriented normal to the buoyancy force vector or horizontally, as shown in Figure 10.3. Any oil in this environment moves from higher toward lower levels oil potential energy illustrated as U_O. The isopotential energy planes for gas (U_G) are oriented similarly, although the buoyancy force upward is stronger than for oil, as reflected by the longer length of the white arrow in Figure 10.4. The potential energy minimum for all hydrocarbons is at the highest point in any hydrologically static structure, and entrapment requires that this point be completely bounded on all sides by reservoir walls or similar hydrocarbon-resistant barriers. Thus, in a static reservoir the U, V, and Z traces are level, and parallel to each other, as are hydrocarbon-water contacts.

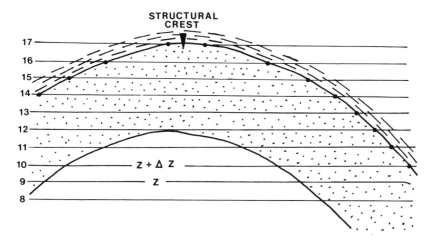

FIGURE 10.1. Cross section of a simple anticlinal structural trap showing traces of Z planes, which are horizontal traces of reference contour elevations.

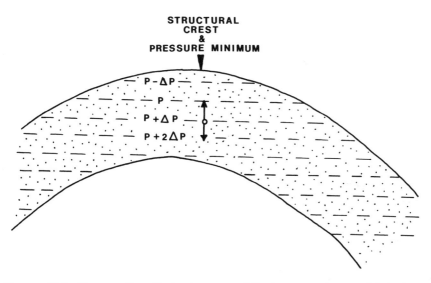

FIGURE 10.2. Cross section illustrating an anticlinal structural trap with hydrostatic reservoir in which forces acting on the water are in balance and pressure gradient is vertical, with isopressure planes horizontal. Hydraulic head, H_W, is constant under static conditions.

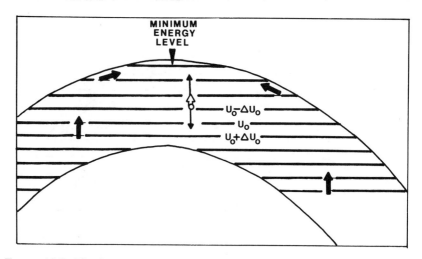

FIGURE 10.3. The forces acting on elements of oil in the static structure shown in Figure 10.2 are not in balance; buoyancy upward creates the potential energy gradient with isopotential energy planes (U_O) oriented horizontally. Black arrows represent general migration directions.

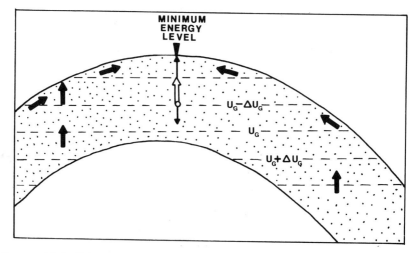

FIGURE 10.4. Orientation of isopotential energy planes for gas (U_G) is also horizontal within a hydrologically static reservoir. Force direction are the same as for oil, but magnitudes are greater.

Hydrodynamic Structural Trap

The forms and positions of both oil and gas traps as affected by moving formation water can be modeled to predict pool locations and their maximum possible size and whether the structure in question can indeed trap and hold hydrocarbons under varying hydrodynamic reservoir conditions. The U, V, Z procedure as outlined in Figure 9.14, is utilized for discerning the orientation of the hydrocarbon isopotential surfaces (U_O and U_G) under specified sets of geologic and hydrologic conditions, and the results show when entrapment is either feasible or highly unlikely.

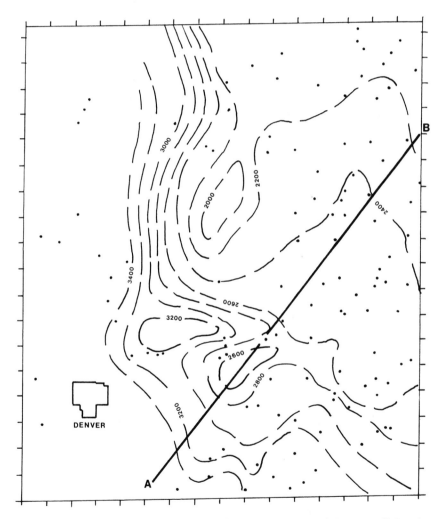

FIGURE 10.5. Potentiometric map covering an area east of Denver, Colorado, constructed from pressures measured in the D&J Sand. (Published by R. Hoeger.)

The source of the required hydrologic parameter, V, is an appropriate potentiometric map, which essentially depicts geographic variation in the hydraulic head (H_W). V then equals TAF × H_W. As an example, Figure 10.5 shows a regional potentiometric map covering an area to the northeast of Denver, Colorado. The hydraulic head contours suggest a general flow gradient from west to east, with several local "highs" and "lows". Figure 10.6 shows how along the line of section indicated in Figure 10.5 the potentiometric elevations read off the map are converted to V_O's and then projected into the plane of the cross section as labeled traces mimicking as realistically as possible the presumed flow pattern.

The case of a hydrodynamic reservoir is illustrated in cross section in Figure 10.7, which depicts formation water moving from left to right through the structure, as indicated by the white arrows. The broken lines are the traces of the water isopotential energy surfaces with respect to oil (V_O), which show the loss of potential energy in the direction of flow. The hydrodynamic gradient is mirrored by the sloping trace of the potentiometric surface in the direction of flow. The orientation of the U_O surfaces can be determined by overlaying the Z traces in Figure 10.1 with

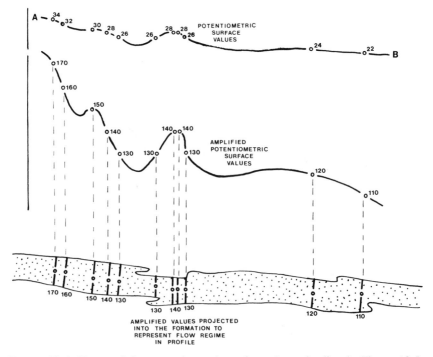

FIGURE 10.6. Trace of the potentiometric surface along the line in Figure 10.5, showing the trace of the water isopotential surface (V_O) and the corresponding V_O planes projected down into the cross section of the formation.

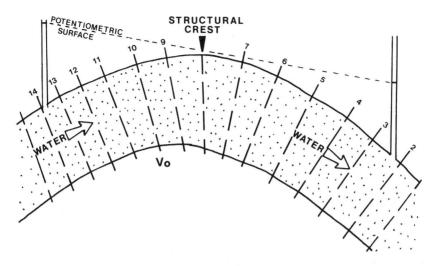

FIGURE 10.7. Cross section under hydrodynamic conditions with water moving from left to right, showing traces of water isopotential surfaces (V_O) as related to oil.

the V_O traces in this figure and subtracting the former from the latter at the line intersections. The result then is the two-dimensional grid of U_O values shown in Figure 10.8. From this grid, the traces of the U_O surfaces can be sketched, as in Figure 10.9. This diagram is essentially a two-

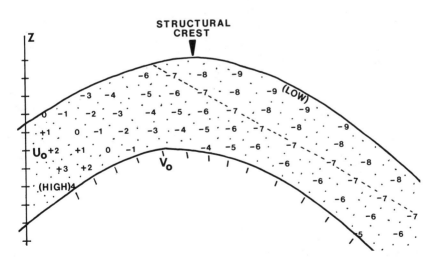

FIGURE 10.8. Oil isopotential energy values (U_O) calculated as $U_O = V_O - Z$ at grid point locations within the plane of the cross section.

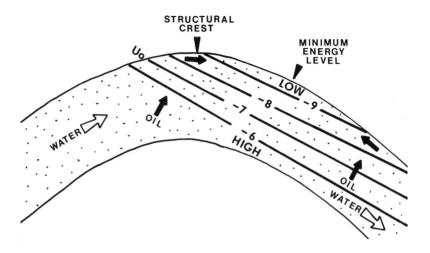

FIGURE 10.9. Traces of the oil isopotential energy traces (U_O) constructed from the grid values in Figure 10.8. Forces acting on elements of oil are oriented normal to these surface traces.

dimensional map showing the levels of potential energy for any oil at any point throughout the water-saturated hydrodynamic aquifer. If droplets of oil were released within this system in such a way that we could witness their motions, we would see that they follow paths influenced by the forces oriented similar to the black arrows in the diagram! The very first droplet would arrive at the point labeled "minimum energy level," and as more and more followed, the pool would build in such a way that at any moment in time the oil–water interface would parallel the oil isopotential energy traces (the U_O lines), which are normal to the force direction. Once trapped, the oil in the pool becomes static and the U_O's vanish since there are no potential energy gradients within a static body of any fluid.

For gas in the same hydrologic environment, the entrapment potential cross section illustrating this scenario is constructed in the same way. The V_O's are converted to V_G's because of the density difference between oil and gas. As illustrated in Figure 10.10, the water isopotential energy surface traces V_G relative to gas are more widely spaced than the V_O traces because for oil,

$$V_O = \frac{D_W}{D_W - D_O} \times H_W, \qquad [10.1]$$

in which D_W = density of water, D_O = density of oil, and D_G = density of gas.

For gas in the same water flow regime,

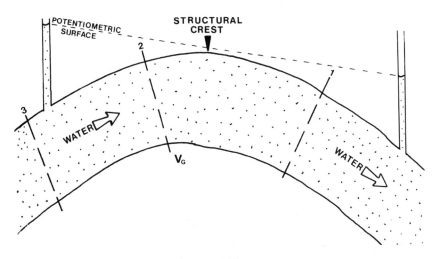

FIGURE 10.10. Traces of the water isopotential energy planes (V_G) with respect to gas, calculated from hydraulic head values and the potentiometric surface.

$$V_G = \frac{D_W}{D_W - D_G} \times H_W. \qquad [10.2]$$

In this example, if $D_W = 1.0 \, \text{gm/cc}$, $D_G = 0.4 \, \text{gm/cc}$, and $D_O = 0.87 \, \text{gm/cc}$,

$$V_G = 1.66 \, H_W \qquad [10.3]$$

and

$$V_O = 7.69 \, H_W. \qquad [10.4]$$

The water isopotentials (V_O and V_G) relative to oil and gas phases are functions of hydraulic head, H_W, at any point.

To express V_O in terms of V_G,

$$V_G = 1.66 \, H_W, \qquad [10.5]$$

so that

$$H_W = \frac{V_G}{1.66}. \qquad [10.6]$$

Substituting in [10.4] gives

$$V_O = 7.69 \, \frac{V_G}{1.66} = 4.6 \, V_G, \qquad [10.7]$$

or, V_O is about five times V_G. Thus, the loss of a single unit of potential energy by the water relative to gas will take place over five times the flow distance relative to oil, or one-fifth as rapidly over a constant distance.

Figure 10.10 shows the traces of the water isopotential surfaces with respect to gas in the same hydrodynamic environment (ΔH_W is similar in both cases). Calculating the grid of U_G values ($V_G - Z$) and contouring yields the family of U_G surface traces in Figure 10.11. These describe the forces that control the movement and entrapment of gas under the conditions in this structure. The minimum energy level point for gas is slightly displaced in the direction of water flow, as illustrated, and any gas–water interface will parallel the U_G traces.

With entrapment of both fluids, the positioning of the hydrocarbons would resemble that in Figure 10.12. The gas occupies the upper position because of its superior resultant force vector. The oil is displaced in the downflow direction, as shown. In a geologic trap, less dense fluids displace more dense fluids. Thus, oil will move water, and if gas finds its way into the oil-saturated structure it will push the oil downward toward any spill point, where closure dies out. Below this point, the oil escapes, and once the entire structure is gas saturated, no other more dense incoming fluid can displace it again until gas is removed to make new room. These mechanics underlie the ingenious concept of "differential entrapment" (Gussow, 1963). The only possible way for a trap to be "flushed" of hydrocarbons is by moving water; thus, it can occur only in a hydro-dynamic environment. Since the gas and oil, *once they are trapped*, are essentially physically static phases of differing density, the interface between them is *horizontal*, as illustrated in the figure, even though the hydraulic environment in the overall formation is dynamic. For dips on

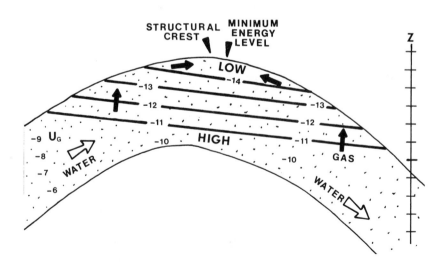

FIGURE 10.11. The family of U_G gas isopotential energy surface traces that reflect the orientation of the principal force (normal to these planes) acting on any gas under these conditions.

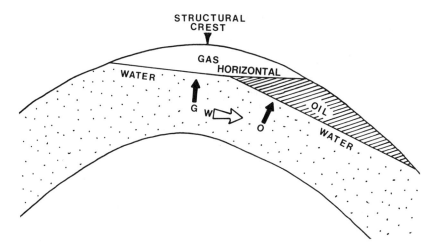

FIGURE 10.12. Configuration of oil and gas pools in this anticlinal structure produced by water flow through the aquifer, as indicated by the hydrodynamic gradient from left to right across the structure.

oil–water or gas–water contacts to occur, the water (the denser phase) has to be moving beneath and in contact with the oil or gas (the less dense phases).

Could this simple structure trap gas or oil if the hydraulic gradient were doubled? By "hydraulic gradient" we mean the rate at which the moving fluid is losing its energy as a result of friction as it flows within the pore spaces of the reservoir rock. Because of its greater buoyancy, gas resists the forces of the moving water more strongly than oil, but just how much more must be determined by employing the U, V, Z cross section construction procedure. The result is shown in Figure 10.13. The gas isopotentials are tilted more steeply than before (which is predictable) but not to the extent that they exceed the dip on the downflow side of the structure. The tilt on the oil isopotentials is, however, greater than the structural dip in this case, so no oil can be trapped under these conditions (i.e., the water flow forces are significantly greater than the oil buoyancy force combined with the containment ability of the structure). The net force component, represented by the large arrow, is resolved in a downward direction parallel to the top of the reservoir. The result indicates that this feature can hold gas but is flushed of its oil in this hydrodynamic environment.

The preceding cases show the U, V, Z procedure applied to problems of defining the orientation of the hydrocarbon isopotential surfaces. This is done to test the possibility of oil and gas traps existing as a function of the hydrologic and structural geologic components of the subsurface environment.

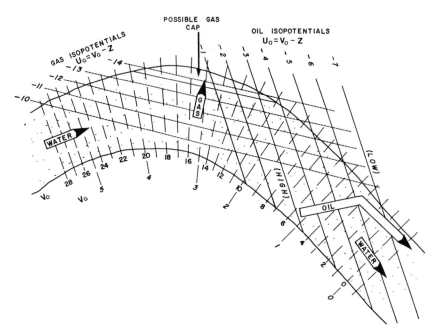

FIGURE 10.13. Cross section showing the entrapment picture for oil and gas in the same structure but with a hydrodynamic gradient that is twice that tested in the previous case.

Stratigraphic-Hydrodynamic Trap Model

In a stratigraphic trap, oil and gas entrapment is primarily a function of internal lithology changes and hydrologic environment and only minimally of structure. Hydrocarbon accumulations can occur when and where the combined effects of these factors are optimal.

A cross section through a uniformly dipping sandstone unit with reference elevation lines (Z) superimposed over it is shown in Figure 10.14. For this aquifer to retain hydrocarbons, some internal factor is required to create a warp in the oil isopotential surface traces so that they close against the homoclinally dipping roof of the reservoir. Certain variations in flow intensity caused by porosity and permeability changes can produce this effect. Figure 10.15 illustrates a scenario in which the downdip portion is significantly lower in permeability so that the water movement is markedly diminished because of increased rate of energy loss from pore wall friction. The comparative slopes in the two different portions of the potentiometric surface reflect the tighter lithology, as does the diminished spacing of the water isopotential energy surface traces (V_O), which show that the water loses "one unit" of energy over a

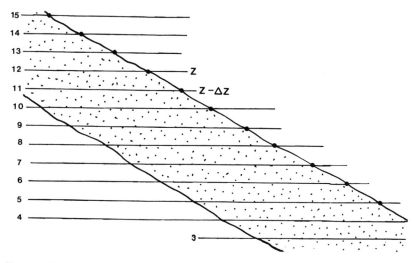

FIGURE 10.14. A homoclinally dipping aquifer showing the configuration of the reference elevations (traces of the Z planes) in the cross section.

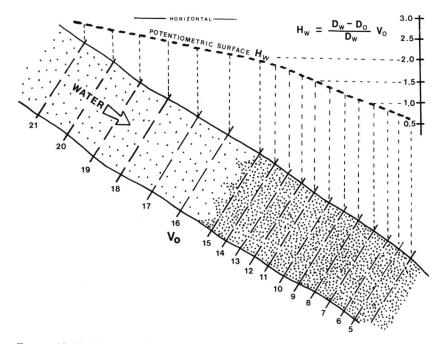

$$H_w = \frac{D_w - D_o}{D_w} \, V_o$$

FIGURE 10.15. Traces of the water isopotential surfaces (V_O), indicating downdip flow through the aquifer unit and trace of the local potentiometric surface (H_W) reflecting the hydrodynamic gradient from left to right.

shorter distance within the tight part of the aquifer than in the open flow section updip. The traces of the corresponding oil isopotentials, calculated as $U_O = V_O - Z$, plunge toward the bottom of the aquifer within the semipermeable zone, as shown in Figure 10.16.

Figure 10.17 shows the configurations of the potentiometric surface and related amplified H_W levels, which are the traces of the V_O water isopotentials with respect to oil. The influence of the zone on the flow pattern is reflected by the "break" in the potentiometric surface.

The traces of the related oil isopotential surfaces constructed using the U, V, Z procedure are shown in Figure 10.18. This is one of Hubbert's original eye-opening cases. The hydrodynamic downwarping of the U_O traces creates a potential energy low with respect to oil. A small body of oil in this section would be moved in response to forces oriented normal to these traces from higher levels toward lower levels. It is worth noting that formation tests in wells located at A and B would both recover water, although the pressures (and potential energy levels calculated from them) would differ from each other, indicating flow. Applying the U, V, Z construction method on a regional scale highlights the probable locations of oil and gas traps created by favorable combinations of geology and fluid mechanics.

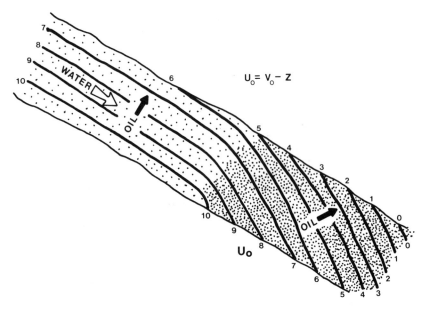

FIGURE 10.16. The traces of the oil potential energy surfaces (U_O) calculated as $V_O - Z$ on the upflow side of an internal semipermeable barrier.

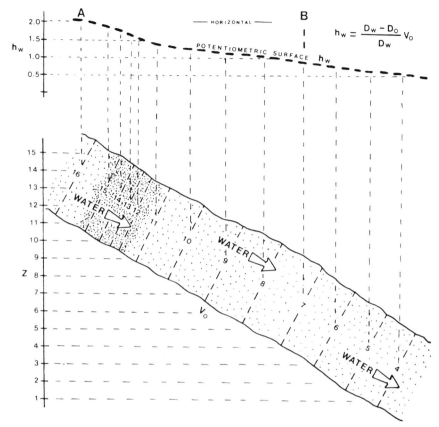

FIGURE 10.17. Trace of the potentiometric surface and internal V_O planes on the downflow side of the semipermeable barrier in a hydrodynamic, homoclinally dipping aquifer.

Regional-Scale Entrapment Potential Cross Sections

The following examples demonstrate how entrapment potential cross sections apply to exploration problems for identifying hydrodynamically favorable and nonfavorable locations for accumulating pools of oil and gas.

Figure 10.19 shows a potentially hydrocarbon-bearing unit in cross section. The trace of the potentiometric surface indicates that the unit is hydrodynamic, with water flow from east to west. The only closed structure is A, and the size of the largest possible hydrocarbon accumulation that it could hold under static conditions is proportional to the cross-hatched area in the section. This entrapment potential cross section is

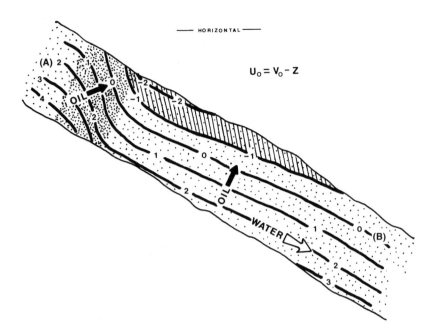

$$U_O = V_O - Z$$

FIGURE 10.18. Orientation of the oil isopotential energy surface traces (U_O) on the downflow side of the internal barrier, indicating potential for an oil accumulation. Within the barrier, water should flow faster as described by Bernoulli's Principle.

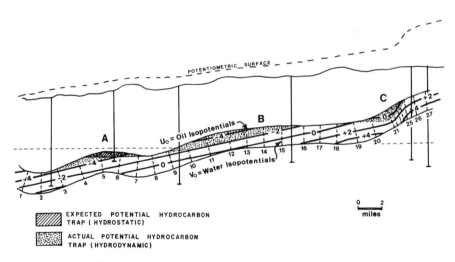

FIGURE 10.19. Regional-scale entrapment potential cross section constructed using the U, V, Z procedure, which discloses in this case that the hydrodynamic influence is constructive, creating one major trap and increasing the hydrocarbon-holding capacity of another.

constructed using the U, V, Z method showing downdip flow with the V_O surface traces decreasing from 27 down to 1 across the length of section, reflecting the potential energy loss in the water along the flow path. It demonstrates that in this hydrodynamic system the anticipated oil pool in the structure would be slightly displaced in the downdip direction, and the volume of oil could be larger than under static conditions by virtue of the dipping oil–water contact, which increases the trap size. In addition to this enhanced entrapment, the dynamic nature of the reservoir creates the possibility of an even larger pool near B. The oil–water transition zone at any moment in time will be parallel to the heavy U_O oil potential energy surface traces. All that is required is that the oil arrive in this trap while these conditions exist. The downwarping of the U_O traces in the vicinity of C discloses the possibility of a facies-barrier oil accumulation at this location. None of these clues would be apparent if a hydrostatic environment were inferred, and an important oil pool could be over-looked without the U, V, Z tip-off to locations where the hydrodynamic factor is constructive.

In the scenario illustrated in Figure 10.20, the hydrogeology is destruc-tive to entrapment, but this would not be known, prior to the drilling of needless dry holes, without the modelling the section. The large mapped structure at location A is a most tempting target since it is closed with a potential hydrocarbon capacity proportional to the cross-hatched area in the diagram. That is how much oil or gas the feature can hold under hydrostatic conditions! But as the potentiometric surface trace shows, the

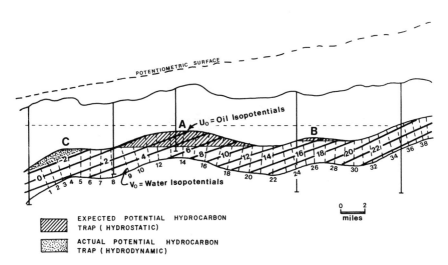

FIGURE 10.20. Regional-scale entrapment potential model showing destructive downdip flow resulting in flushing of the two closed structures with migration into a nonclosed nose downdip.

conditions are dynamic with a hydraulic head gradient from east (high) to west (low), reflecting water flow downdip in that direction. The U_O traces tilt in the downflow direction as well, and the net force acting on any oil in this stretch of reservoir is normal to these traces, or up and to the left in the figure. Since the tilt on these surfaces exceeds the dip on the downflow side of structure A, and the oil–water contact at any moment in time has to parallel these traces, no trap exists in this feature; it is essentially flushed. The entrapment potential analysis also reveals that some of the displaced oil is retrapped in the nonclosed, downdip nose on the left end of the cross section.

The cross section in Figure 10.21 shows the traps that exist under static conditions where buoyancy is the major force operating on the hydro-carbons in the reservoir, along with those created by hydrodynamic conditions where a combination of water flow *and* buoyancy is the major force. As the U, V, Z construction discloses, the intensity of the downdip flow reflected by the westward-sloping potentiometric surface drastically diminishes the entrapment potential of the feature. Under static condi-tions, the oil pool could be enormous, but U, V, Z analysis indicates that the hydrodynamic effect has flushed most of the structure. Two smaller pools are possible by consequence of the downdip water flow, but these

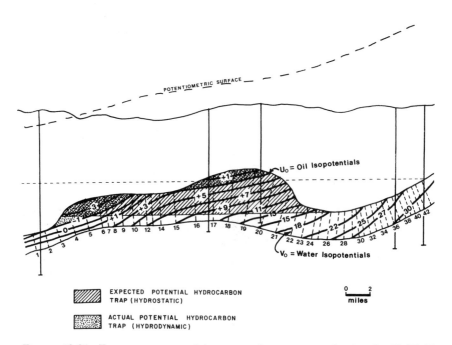

FIGURE 10.21. Entrapment potential cross section constructed using the U, V, Z method, which discloses that under the conditions illustrated, the capacity of the large structure would be severely diminished by downdip flow at the scale shown.

are minimal in comparison with the accumulation that the large central structure could harbor if the hydraulic environment were benign.

Figure 10.22 shows entrapment potential cross sections for a single aquifer unit but for cases of both downdip and updip formation water flow. The figure discloses that under static conditions a trap at location C would hold a small accumulation of oil, controlled by the spill point in the structural low between C and D. In the downdip case, as the U_O traces reflect, traps are created at locations A, D, and E, and the pool at C is enlarged by virtue of displacement downdip toward B resulting from downward-flowing formation water. When the flow is reversed from downdip to updip, all of these traps are eliminated, and the oil force vectors (normal to the U_O traces) drive the hydrocarbons up toward the basin margin. These two entrapment potential models demonstrate again that updip flow is not conducive to the trapping of oil and gas, while downdip flow creates and enlarges hydrocarbon accumulations.

Entrapment potential cross sections such as these provide the exploration geologist with a procedure for assessing the favorability of trapping conditions in any geologic feature under any set of hydrologic conditions deemed realistic. The success of the results in terms of predicting reserves

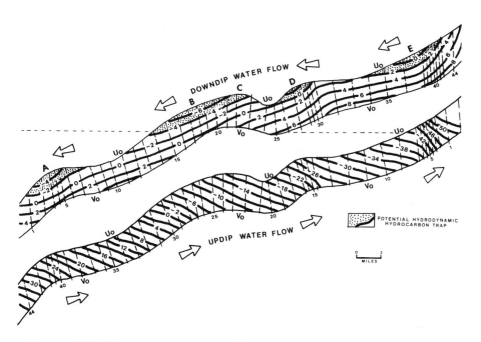

FIGURE 10.22. Two entrapment potential cross section models illustrating the constructive effects of downdip flow in a dynamic aquifer with the nonconstructive effects of updip flow in the same unit.

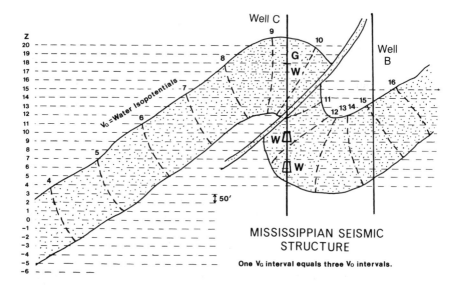

FIGURE 10.23. An entrapment potential cross section problem for the reader to solve.

depends on how accurately the investigator is able to mimic the actual reservoir geometry, the lithologic variability within the rock itself, and the hydrologic conditions.

U, V, Z Oil Prediction Problem

The structure in the cross section in Figure 10.23 is provided as a practice problem for the reader in U, V, Z entrapment potential cross section construction. The apparently tilted gas–water contact suggests (if the log picks are accurate) the possibility of water flow from left to right. The appropriate traces of the V_G planes have been sketched in. The gas is very "wet," which, in combination with the oil shows in the water recovered from well C, suggests the possibility of free oil nearby. Using the U, V, Z procedure, one can test whether the environment supports an oil leg downdip from well C. The following steps should be followed:

1. Sketch traces of the water isopotentials with respect to oil (V_O), by subdividing the present V_G spacing into three V_O intervals. Assign new numbers to the V_O traces, say from 1 up to 25.
2. The Z traces are projected across the section as the broken lines.
3. At each V_O and Z intersection, compute and post the appropriate U_O values.

4. Contour the resulting grid of U_O values to define the oil potential energy field within the reservoir. Show with arrows the orientation of forces that would be influencing the movement of oil.
5. Draw and label the oil–water contact that would bound the largest possible oil reservoir.

11

Entrapment Potential Maps

Hydrocarbon Entrapment

The trapping of hydrocarbons occurs in subsurface geologic features when the orientation of forces is such that the oil and gas are held in place and cannot escape in any direction. One classification scheme, devised by Meissner (1984) and illustrated in Figure 11.1, proposes a three end-member system constituting a triangle. The three end-members in the scheme occupy positions at its apices: *structural traps* (lower left), *stratigraphic traps* (lower right), and *hydrodynamic traps* (top center). Structural traps are geometrically "closed" on all four sides (two sides in cross section) and the hydrocarbon–water contacts are horizontal. The reservoirs are hydrostatic. The influence of the structural component decreases upward along the left leg of the triangle. The first dynamic pool is an accumulation trapped off the crest of the structure, held in a downflank position, with a dipping oil–water contact. The next feature is a monocline, also nonclosed with a tilted oil–water contact. The last pool on this leg, close to the hydrodynamic end-member position, is essentially a slug of oil localized at the bottom of a synclinal trough, held there by convergent downdip flow. No one, to our knowledge, has ever described an oil accumulation of this kind, but theory and hydromechanics suggest that it could exist under proper hydrodynamic conditions.

In the same position on the right-hand leg of the classification triangle is a completely structureless accumulation positioned in a horizontal reservoir. The mechanism is converging flow which, while the water escapes downward along the fault plane, the oil cannot and is held in place as shown. Whether such accumulations really exist is unknown! In the next case, unidirectional flow with a local barrier creates the trap for the oil accumulation. The remaining two traps are common unconformity and reef-type traps, primarily stratigraphic, as is the updip pinchout feature along the bottom leg of the classification triangle.

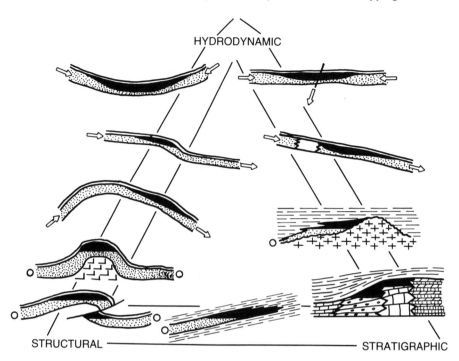

FIGURE 11.1. General hydrocarbon trap classification, showing three end-members: structural, stratigraphic, and Hydrodynamic. (From F. Meissner.)

Hydrodynamic-Structural Trap and Entrapment Potential Mapping

The U, V, Z procedure is utilized for constructing maps that indicate the hydrocarbon entrapment potential of both local and regional structural and stratigraphic features with respect to the relevant hydrologic conditions. The resulting maps demonstrate possibilities in a given area where the hydraulic effects, the geologic framework, the lithologic factors, and the physical attributes of the fluids themselves combine favorably. These are used along with other subsurface maps for spotting optimal drilling locations. On a map, the traces of the U, V, and Z planes are represented as contour lines.

The structure illustrated in Figure 11.2 by the Z elevation contours is a narrow nose, closed on the west, south, and east sides but open to the north. Under static conditions there could be no oil and gas entrapment since the buoyancy of any hydrocarbons would carry them out of the structure and on updip. The V_O lines (the tilt-amplified potentiometric surface hydraulic head values) reflect water flow from north to south. The

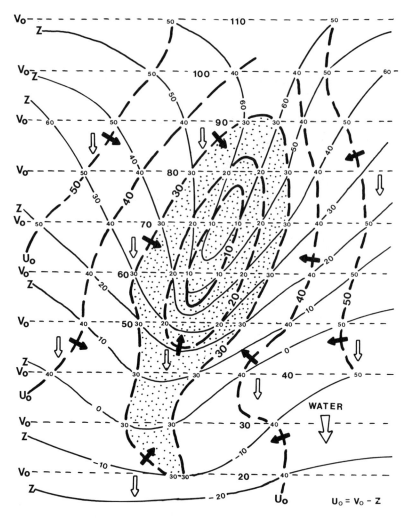

FIGURE 11.2. Entrapment potential map showing a possible oil trap created by hydrodynamic and structural factors in a nonclosed nose structure.

numbers are scaled integers; in the real case they would be coded subsea elevations.

The ability of the structure to trap hydrocarbons under these conditions is reflected by the U_O oil isopotential energy contours, since the net forces behind the oil movement in the reservoir will be oriented normal to these lines, as shown by the black arrows. The potential energy minimum, as the U_O contour pattern discloses, is at the lowest point on this surface; all oil in the vicinity will migrate toward and accumulate at this point. The accumulation will grow from the minimum point outward, the pool boundary at any moment in time will be close to parallel to the U_O contours, and the size of the largest pool possible under these conditions will be proportional to the area enclosed by the *highest* closed contour. In this case it is the U_O line labeled "30." The U_O contours are as before calculated as $V_O - Z$ at the contour intersections, or at interpolated points over a grid if better resolution is required and computer equipment is available.

The shaded portion of the map, enclosed by the heavy U_O contours, indicates that the hydrodynamic conditions are favorable for oil entrapment in this structure; potential for a large pool is also indicated. The flow toward the south, the buoyancy force of the oil directed toward the north, and the containment provided by the sides of the structure are thus all serving in concert to create an oil trap here.

Whether the same structure can trap gas under these conditions can be tested similarly. The only change necessary is that the V_O's be replaced by V_G's, as shown in Figure 11.3. A V_G/V_O ratio of 1/5 has been used to transform the V contours as shown. The resulting U_G map represented by the heavy contours indicates that this structure can hold a miniscule amount of gas but that the overwhelming buoyancy force of the gas will cause most of it to move up and out of the open updip end of the structure. The black arrows indicate the orientation of the gas force vectors as a function of buoyancy, hydrodynamic force, and structural containment.

The reader who is curious as to whether the structure could trap gas if the hydraulic gradient were twice as intense can test this possibility using the U, V, Z procedure and the map in Figure 11.4. In this figure, the V_O's have to be converted again to V_G's, and U_G must be calculated and posted at the V and Z intersections and then contoured.

Entrapment in such situations is essentially a tug-of-war between the natural buoyancy of the hydrocarbons upward and the flow force of the dynamic formation water in a different direction, plus whatever natural closure is provided geologically by the walls of the reservoir itself. The U, V, Z method merely summarizes the combined effects of these factors. The end-member cases are flushing hydrocarbons from the structure by dominant water flow when the hydrodynamic component is too strong; and updip migration and loss of hydrocarbons when the downdip hydrodynamic component is too weak.

Displacement of Hydrocarbon Pools

In some naturally hydrodynamic units, the magnitude of the water movement is so extreme that an individual hydrocarbon pool is not only tilted within the reservoir but actually moved out of it from its original position (fixed by closure geometry) to a new location. Intense hydrodynamic

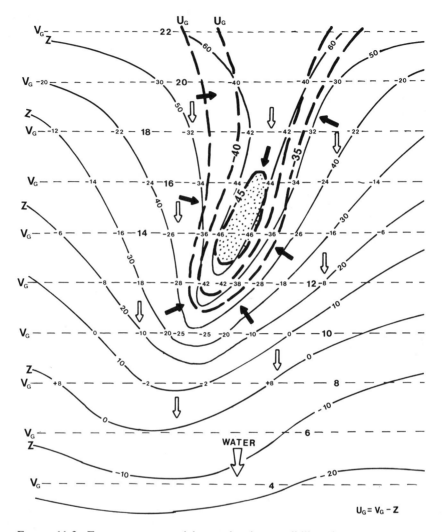

FIGURE 11.3. Entrapment potential map showing possibility of gas accumulations in the same hydrodynamic structure as in Figure 11.2.

conditions remove and dissipate the hydrocarbons, leaving residual oil-staining in water-saturated reservoir rock to reveal that flushing has occurred and the oil has remigrated. The U, V, Z method is effective for predicting, prior to drilling, whether significant geographic displacements have occurred in areas of exploration interest.

The Coles Levee field of California is an example of hydrocarbon displacement in a drilled hydrodynamic area, described by Davis (1952). Figure 11.5 shows two separate pools, both of which have been displaced over 1,000 ft down their respective structures in an easterly direction. The corresponding dip on the oil–water contact is around 45 ft/mi to the east. The cross section shows the tilted oil–water contact and gas–oil contact as well.

The Sage Creek, Wyoming, pools shown in Figure 11.6 constitute a famous example from the Bighorn Basin. This example especially illustrates the importance of entrapment potential mapping to exploration projects. An interpretation of the feature (contours on the Tensleep

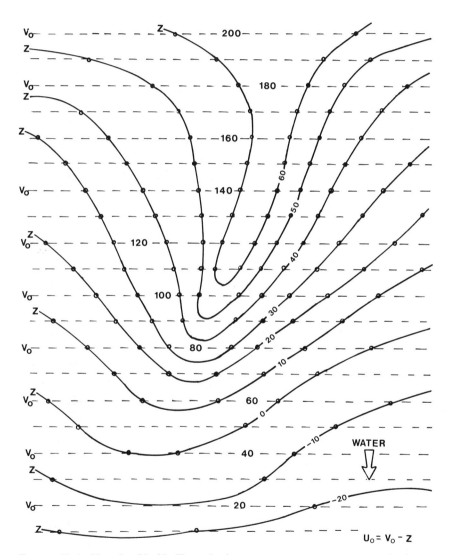

FIGURE 11.4. Use the U, V, Z method to construct a map to test whether a potential oil and gas trap could exist at this location as a result of local structural and subsurface flow conditions.

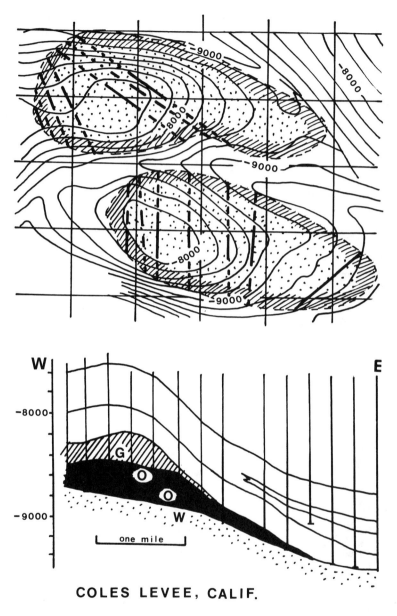

COLES LEVEE, CALIF.

1. Oil leg displaced downdip
   to the east from structural
   crests.

FIGURE 11.5. Structure contour map showing outline of two oil pools and cross section through the Coles Levee oil field. (From Davis, 1952.)

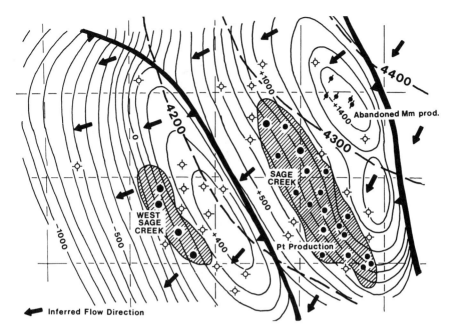

FIGURE 11.6. Structure contours and pool outline of the Sage Greek and West Sage Creek oil pools in the Bighorn Basin of Wyoming. (After Pedry, 1975.)

Formation) shows an anticline with about 400 ft of structural closure bounded to the east by a major fault. Major production is from two sands, and the largest 8,000,000-barrel oil pool shown in the figure is displaced to the southwest of the crests, to the extent that wells drilled within the top three closing contours would miss the production. Also significant is that one sand in the three updip wells is stained with live oil and no tar, suggesting hydrodynamic displacement. Also, the dip on the oil–water contact is close to 800 ft/mi, and the bottom water is extremely fresh. The potentiometric surface contours for the Tensleep Sandstone imply several hundred feet of hydraulic head drop from northeast to southwest. The same displaced-pool scenario is repeated to the west in the West Sage Creek pool, which is smaller but analogous to the main Sage Creek oil accumulation.

The Upper Valley pool from Garfield County, Utah, illustrated in Figure 11.7, displays similar hydrodynamic displacement. Sharp (1976) describes it as "a structural trap which has been modified by hydrodynamics and stratigraphy." The pool outline reflects its position south and slightly west of the structural crest, with an associated oil–water contact that dips to the southwest. The oil gravity is reported as increasing uniformly from 26° in the highest portion of the pool to 19° on the lower flank of the anticline. An interesting problem for the reader to consider is

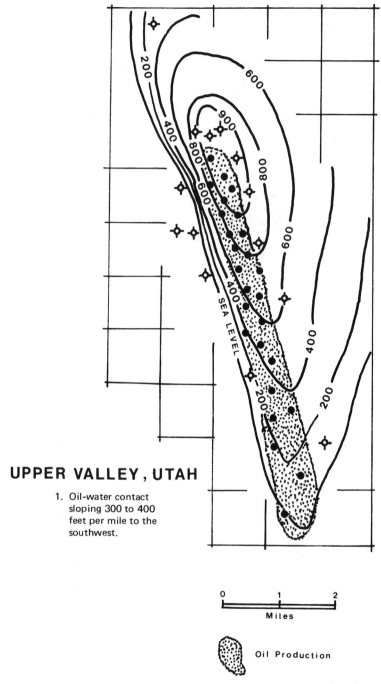

UPPER VALLEY, UTAH

1. Oil-water contact
   sloping 300 to 400
   feet per mile to the
   southwest.

FIGURE 11.7. Subsurface map of the Upper Valley pool, in the Kaiparowits Basin in Garfield County, south central Utah.

that of using the U, V, Z procedure to map the appropriate water flow pattern (reflected by H_W) for the area from the structure contours (Z) and the pool boundary which is a U_O line plus construction of the regional V_O contours. The result may indicate the potential for significant production farther to the southwest. (See Goolsby et al., 1988; Dahlberg, 1993.)

This procedure allows the determination of the range of hydrodynamic conditions (in terms of water flow direction, intensity, and density of the hydrocarbons and the formation water) under which a given structure mapped from subsurface well or seismic data can hold significant (economic) quantities of oil or gas. If the optimal conditions are within the range of conditions appropriate for the particular reservoir concerned, then the prospect may be attractive as a money maker. On the other hand, if the optimal conditions required for a given structure to hold hydrocarbons lie far outside the range of realisic conditions, then the potential is less.

Hydrodynamic-Stratigraphic Entrapment Potential

Hydrocarbon entrapment can also take place in geologic features with minimal structure in which case oil accumulation requires a favorable combination of internal permeability variation and hydrology. Loci for favorable entrapment then occur where irregularities in the water potential energy surface (which reflect facies-related flow pattern variations) intersect the relatively featureless structural surface. To create these "hot spots," highly heterogeneous internal flow conditions must exist to produce significant anomalies in the hydrocarbon potential energy surface (U contours), which indicate where oil and gas trapping conditions are optimal. These appear as $V - Z$ minima.

A simple example of U, V, Z treatment of a stratigraphic entrapment problem is illustrated in Figures 11.8, 11.9, and 11.10. Figure 11.8 shows the structure (Z contours) on the top of a reservoir uniᴛ that is a featureless formation dipping southwest. The locations and data are posted on the map. Note that the Z contours are coded and one Z interval equals 25 ft. Hidden within the unit are several geologic barriers indicated as "less permeable" zones. In the real-life case, they would not be evident, although one of the wells has touched an edge of the larger barrier, so some small indication at this point might appear in the logs and cuttings from this well. One of the wells is an oil well. The V_O map constructed from the amplified hydraulic head values and potentiometric map (H_W) is shown in Figure 11.9; it provides the key to the problem, since the effects of the permeability barriers on the regional water flow pattern show up as prominent anomalies: a "step" and a "scoop." Overlaying the two maps and subtracting at all intersections the Z values from the V_O values

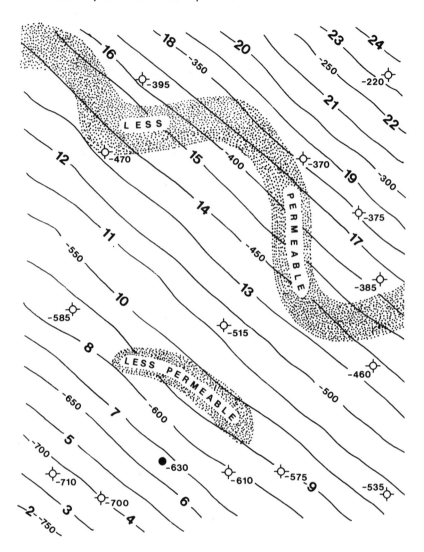

FIGURE 11.8. Structure contour map showing location of hidden internal features, which could influence hydrocarbon entrapment.

produces the isopach between them ($U_O = V_O - Z$) and demonstrates levels of potential energy for any dispersed elements of oil within the water-saturated formation in this area, a combination of moving water and zones of reduced permeability and porosity.

The resulting entrapment potential map is displayed in Figure 11.10. Two favorable locations are indicated as low points on the oil potential energy surface, and the size of the largest possible oil pool in this hydro-facies scenario is proportional to the area enclosed by the *highest* closed contour. For both anomalies this is the "−5" line, and the relevant areas are stippled. This example illustrates that updip water recoveries do not preclude the possibility of hydrocarbon accumulations downdip from them.

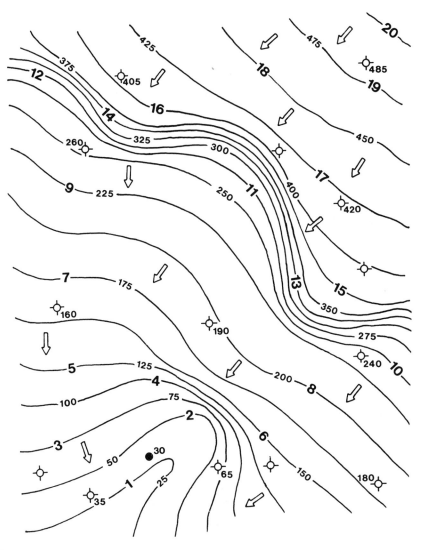

FIGURE 11.9. Amplified potentiometric map (V_O contours) that reflects the effects of the semipermeable barriers in Figure 11.8.

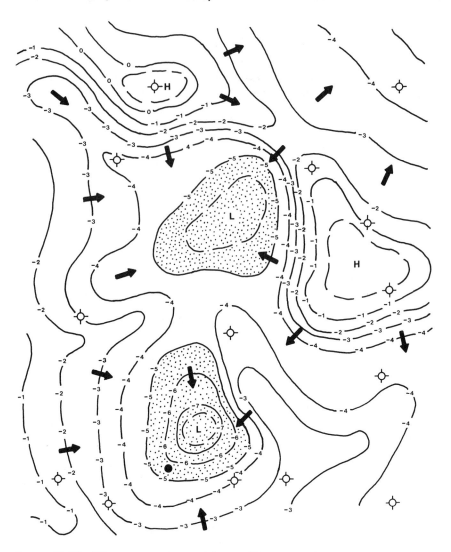

FIGURE 11.10. Oil entrapment potential map (U_O contours) constructed from the data from Figures 11.8 and 11.9.

These models illustrate application of the entrapment potential mapping procedure for predicting locations of traps created by favorable hydrodynamic conditions. They are conceptual and represent ideal conditions in terms of pressure data quality, fluid type, well locations, and identifiable test intervals and formation tops. In reality, of course, subsurface information is rarely complete and numerous assumptions must be made to bridge data gaps. The geologist employing hydrodynamics is faced with the quite usual conundrum that when the factual information becomes adequate, the discovery has by then been made (by someone else), the land is tied up (by other parties), and the prospect is no longer viable or attractive.

The overall success of applied hydrodynamics in petroleum exploration depends strongly on three basic factors. The first involves the quality, abundance, documentation, and geography of the available pressure data from which the potentiometric and entrapment potential vlaues are calculated. The second factor is the appropriateness of the assumptions that underlie the interpretation of the calculated water data and the contouring of the inferred water flow patterns. Hydrodynamic maps are just as interpretive as maps based on subsurface data from other sources.

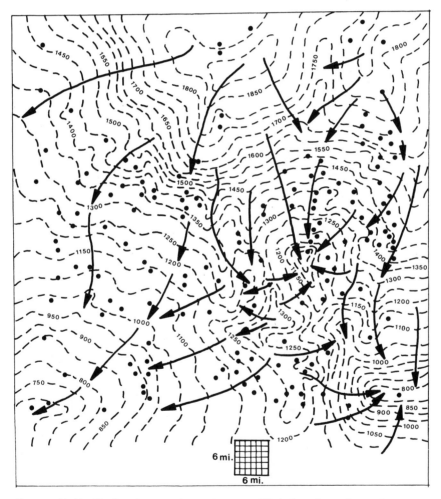

FIGURE 11.11. Regional potentiometric map (H_W) based on formation water pressures measured in the Cretaceous Viking formation in Alberta.

The third factor is skillful interpretation and a generous portion of luck. The ultimate success depends most simply on how well the individual has succeeded in representing either intentionally or by accident, the critical processes in the units concerned.

An Application of U, V, Z Mapping

The problem here is to highlight areas in the Cretaceous Viking formation in a 130-township area in central Alberta, where the potential for oil entrapment appears geologically and hydrologically favorable. The data

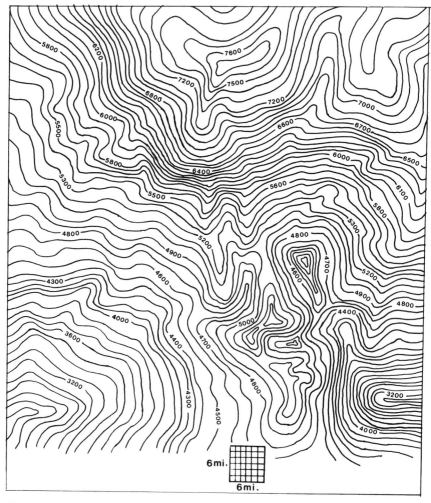

FIGURE 11.12. Regional amplified potentiometric map based on calculated V_O values from Figure 11.11.

consist of pressure and depth (DST) information from 170 wells. Once the formation test values are screened, the calculated hydraulic head values are plotted on the map and contoured to expose the potentiometric surface for the area, as shown in Figure 11.11. This potentiometric map displays a general decrease from northeast to southwest, with a prominent anomaly appearing as a closed low in the east central part of the area. Figure 11.12 shows the V_O water potential surface (relative to oil); this map is thus the amplified potentiometric map from Figure 11.11.

The Z map displaying the structure on top of the Viking reservoir unit is shown in Figure 11.13. The contouring appears conservative since no local structures of any significance show up (at least with a 100-ft contour

FIGURE 11.13. Structure map on the Viking formation. Contours are Z values corrected to a common datum of $-6,000$ ft subsea.

interval); in an area of this size, this is statistically unlikely. Thus, if any oil accumulations are anticipated in this region, the traps will have to be largely stratigraphic/hydrodynamic.

The solution map, which shows the combined effects of all these factors, is the oil entrapment potential map in Figure 11.14. The contour lines are the traces of constructed U_O surfaces, reflecting levels of potential energy for oil throughout the area. There are two local highs and one low apparent on this map, and the overall topography of this surface, from which the directions of oil movement can be inferred, is similar in appearance to the original potentiometric map. This suggests that the flow-

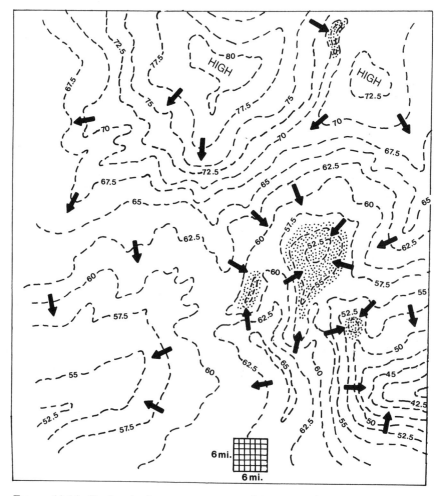

FIGURE 11.14. Regional oil entrapment potential constructed using the U, V, Z method and data from Figures 11.12 and 11.13.

facies components are dominant factors in hydrocarbon entrapment in this relatively structureless unit. On the basis of this map, the explorationist would conclude that factors for oil entrapment are optimal in the vicinity of the closed low seen in the east central portion of the map. Several Viking oil pools are indeed present here.

Remember, the potential for physical entrapment of oil or gas is *high* where the level of potential energy of the oil and/or gas is *low*.

Regional and Local Mapping of Hydrodynamic Traps: Problem 1

This case provides the reader with an opportunity to apply the U, V, Z mapping procedure to a simulated regional-scale problem, where the objective is to recognize hydrogeologically favorable locations for hydrocarbon accumulations. These are exposed by constructing the U_O contours from the available material.

The structure contour map in Figure 11.15 shows contours on the top of the reservoir formation scaled into simple integer values so that the greater values reflect the higher elevations. If one were to predict the most likely locations for petroleum accumulations on the basis of this map only, they would be coincidental with the several identified structures, and these would be good picks *if* the unit is hydrologically static.

The second map (Figure 11.16) has posted on it scattered V_O values (also coded into integers), which are amplified H_W values. The topography of this surface, which is exposed by interpretive contouring of this data, reflects the flow patterns within the reservoir unit (which looks at first glance to be hydrodynamic since the posted head values vary significantly).

To map U_O, which will disclose the orientation of the forces acting on oil present anywhere in this unit, perform the following:

1. On a piece of tracing paper, contour the water potential V_O data on a one-unit interval.

2. Overlay this V_O map on the structure contour (Z) map and compute $U_O = V_O - Z$ at all points where the two contours intersect each other. This is called cross-contouring and is conventionally used for isopaching tops and bottoms of structural data.

3. Contour these difference values (U_O) to produce the oil entrapment potential map (which is really the derived oil potentiometric surface, H_O). Remember, this cannot be calculated directly since we need the potential energy of the oil as a function of the potential energy of the formation water.

Indicate the hydrodynamically favorable locations on the U_O map and compare these closed lows with structurally favorable areas on the Z

FIGURE 11.15. Model structure contour map showing Z elevations on a hypothetical reservoir unit for U, V, Z example.

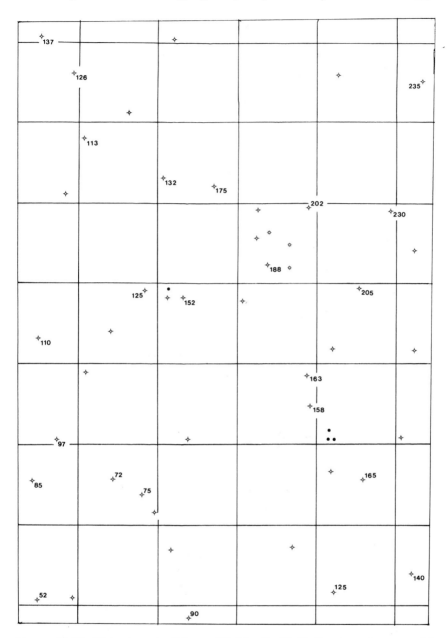

FIGURE 11.16. Water isopotential map (V_O) for the U, V, Z entrapment potential mapping example.

map. The most desirable prospect locations (i.e., areas where the potential for oil trap formation is high) will occur where both the structural and the flow components reinforce each other constructively. The V_O map suggests not only that the unit is hydrodynamic but that there are internal interruptions in the flow pattern. Neither factor is sufficient

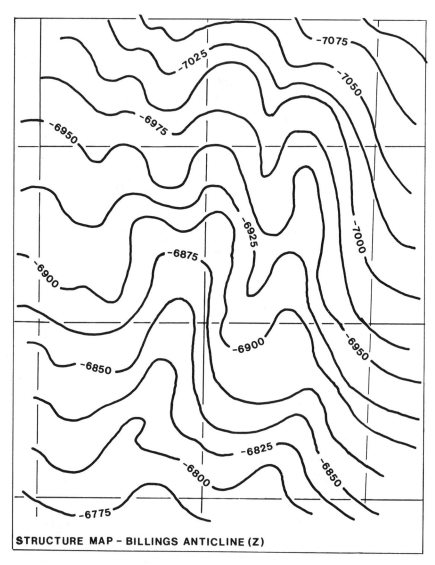

STRUCTURE MAP – BILLINGS ANTICLINE (Z)

FIGURE 11.17. Structure contour map showing a nonclosed northward-plunging anticline in the central Williston Basin of the United States for U, V, Z mapping problem.

alone, so no single map tells the whole story! The acceptable answers to this problem are offered in the Appendix.

Local Mapping of Hydrodynamic Traps: Problem 2

Figure 11.17 is a structure map of a typical plunging anticlinal structure in the Williston Basin (from F. Meissner, after Mitsdarffer, 1985). The contours are subsea elevations of an appropriate gamma ray log marker, which appears to reflect the local structure.

The water flow in the reservoir formation is described as being north-to-northeast in this unit. A potentiometric surface approximation based on this description, the location of oil–water contacts in several marginal wells, appears in Figure 11.18. Assume that this is the best available approximation for the H_W map.

What are the oil entrapment possibilities here? To answer this question, the reader can use the information from Figures 11.17 and 11.18 to construct the U, V, Z map. First, the potentiometric map based on hydraulic head values and reflecting the general regional flow characteristics must be transformed to a V_O water-potential-energy-with-respect-to-oil map by amplifying the contour line values. The tilt amplification factor should be 2.5 in this case.

A transparent overlay can be placed on the potentiometric map and the V_O contours can be drawn. This map is then overlain on the reservoir structure map, and the U_O values calculated as $V_O - Z$ are posted at points where the contours intersect each other. These U_O values are contoured to expose the entrapment potential of this nonclosed structure. The largest possible accumulation under these conditions is proportional to the area enclosed by the numerically *highest* closed contour, and the potential energy minimum for oil is at the lowest elevation within each low. It is important to keep the signs straight in the calculations so that lows are properly low. As an example, $U_O = V_O - Z = -6{,}800 - -(6{,}750) = -6{,}800 + 6{,}750 = -50$, while $-6{,}800 - (-6{,}850) = -6{,}800 + 6{,}850 = +50$. Remember, fluids will move from high levels (maxima) toward local lows (minima). The answer to this problem can be found in the Appendix.

Regional and Local Mapping of Hydrodynamic Traps: Problem 3

In an area where oil production originates from reefal carbonates of Devonian age, combined seismic and subsurface well data indicate the prominent potentially oil-bearing reef structure shown in Figure 11.19.

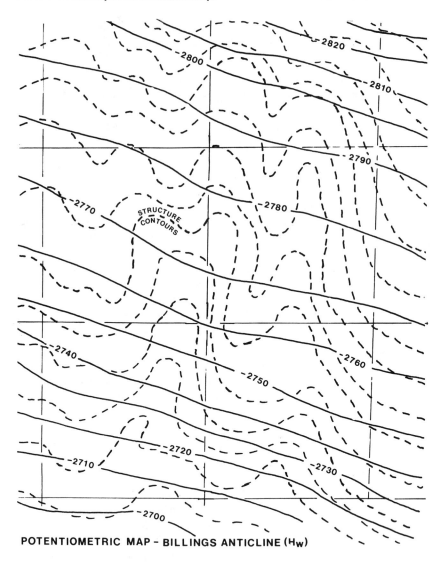

POTENTIOMETRIC MAP – BILLINGS ANTICLINE (H_w)

FIGURE 11.18. Potentiometric map of the Billings anticline region for U, V, Z mapping problem.

Drilling has dashed hopes for a giant accumulation, and although there are significant oil shows, all evidence suggests that the structure has been flushed of its hydrocarbons by strongly north-to-south-flowing formation water. There thus remains the possibility of a displaced oil accumulation, which can be tested by constructing the appropriate oil entrapment potential map. Posted by the well locations in Figure 11.20 are formation

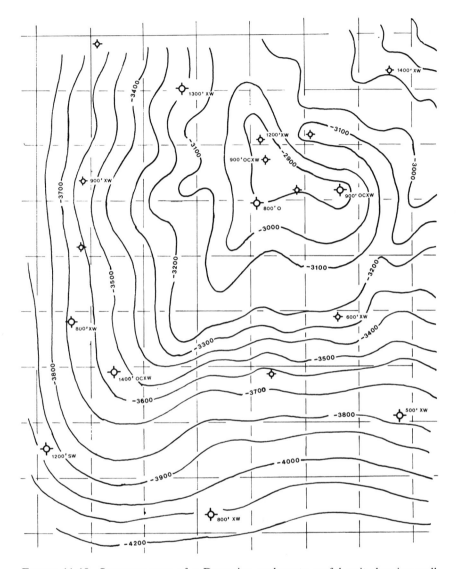

FIGURE 11.19. Structure map of a Devonian carbonate reefal unit showing well locations and DST formation fluid recoveries.

pressures measured in tests in which the water recoveries were significant and the elevation to which the pressure value is assigned.

The first map required is the potentiometric map (H_W) or the water isopotential contour map (V_O), which is made from the posted data in Figure 11.20. The appropriate V_O values can be calculated (assuming a tilt amplification factor value of 5.0) as $V_O = 5(H_W) = 5[KB - RD +$

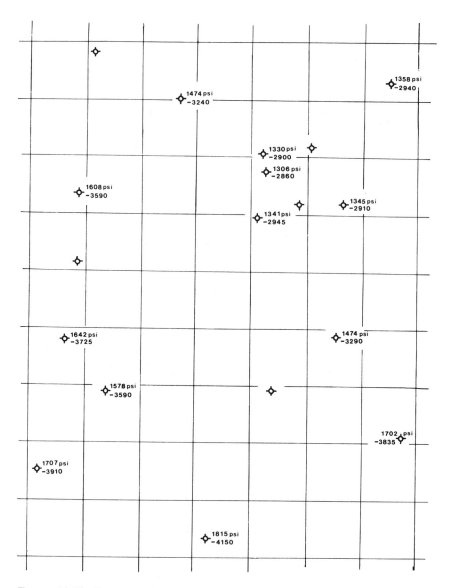

FIGURE 11.20. Base map showing posted DST formation pressures and depth of recorder for U, V, Z mapping problem.

P/0.465]. For the first well, this would be: $5(-3,240\,ft + 1,474\,psi/0.465\,psi/ft) = 5(-3,240 + 3,170) = 5(-70) = -350\,ft$. When these values have been calculated for all the wells and posted on the map, this data can be contoured on tracing paper to yield the required V_O map. The oil entrapment potential map is then completed by overlaying the V_O

map on the Z map in Figure 11.19, calculating the U_O values at the V_O and Z contour line intersections, and contouring the resulting U_O values on the tracing paper. The size of the largest possible accumulation under these conditions will be indicated by the area of closure demonstrated by the numerically highest closed contour line. The result appears in the Appendix.

12

Tilted Oil and Gas Pools:
Some Published Examples

The procedures presented in the previous chapters have been developed from the body of hydrogeological theory that was discussed earlier. It predicts that hydrocarbon–water contacts in oil and gas pools will be oriented nonhorizontally under sufficiently hydrodynamic conditions. The degree of OWC tilt depends on the intensity of flow within the underlying and surrounding formation water, the comparative densities of the moving water, and associated hydrocarbon phases. It was also noted that the methods presented are most effectively utilized prior to drilling, so as to predict whether an accumulation can be trapped and held under existing hydrogeological conditions; if a particular geological feature will be completely devoid (flushed) of oil or gas due to a hostile hydrological environment; whether hydrocarbons might be accumulated in a position offset from the crest of the structure concerned; and the degree of tilting in any pool that might be encountered and how the size of the pool has been affected either constructively or adversely. Several U, V, Z problems in the preceding chapter have illustrated the concepts.

If these ideas have merit and are effective in petroleum exploration and development, there should be reported examples of inclined hydrocarbon–water contacts. There are, and a few are presented below. These are some of the better-known, best-documented cases.

The Wheat Field of West Texas (Adams, 1936) shown in Figure 12.1 consists of oil that has accumulated in a series of nonclosed structural noses within a generally homoclinal sand where, due to lack of four-directional closure, oil would be difficult to trap under nonhydrodynamic conditions. The formation water flows eastward within the reservoir unit, and the oil–water contact dips in the same direction as the theory predicts it should.

Figure 12.2 shows a cross section through the Northwest Lake Creek oil field in Wyoming from Green and Ziemer (1953). This pool vividly demonstrates the prominent effects of northwestward, regional water flow that has pushed and is holding the major oil leg in the Tensleep sand

230

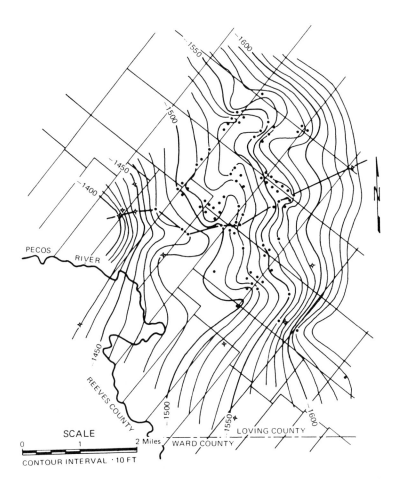

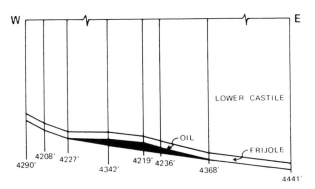

1. Homoclinally dipping sand,
   structural terrace.
2. Tilted oil-water interface.
3. Regional flow to the east.

FIGURE 12.1. Structure contours over and cross section through the Wheat Field in West Texas. (From Adams, 1936.)

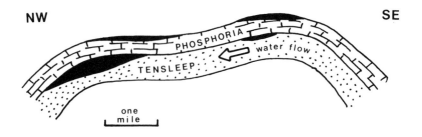

1. Accumulations on NW plunging anticlines.
2. Oil/water contact tilt exceeds plunge.
3. Oil leg offset from crest by 6 miles.

FIGURE 12.2. Cross section through the Northwest Lake Creek oil field in the Big Horn Basin of Wyoming. (After Green and Ziemer, 1953.)

several miles down the nose of the structure. The oil–water contact is strongly tilted in the direction of flow in agreement with the theory for both Tensleep accumulations, as well as the overlying Phosphoria. Drilling at the highest point on this particular structure would yield little petroleum due to its hydrodynamic nature.

The oil–water contact in the South Glenrock, Wyoming pool is inclined to a significant degree, as shown in Figure 12.3. The northeast regional water flow direction is indicated by the pronounced northeasterly slope of the related potentiometric surface. As indicated by Stone and Hoeger (1973), the slope of the oil–water contact observed in the Dakota sandstone and illustrated in the cross section approaches 500 feet per mile in the same direction as the inferred regional water flow.

Strong evidence of the effects of a hydrodynamic reservoir environment is provided by the cross section through the Cushing, Oklahoma field in Figure 12.4 (after Beal, 1917). On the east flank of the structure in the Layton and Wheeler sands there is no oil leg with only gas over water, while on the west flank of the structure there is an oil leg as well. The oil–water contact is likewise inclined to the west, implying an east-to-west hydraulic gradient with the oil having been displaced to the west by the moving formation water. It is difficult to account for this condition by any other feasible mechanism; hydrodynamics explain it easily. Section 32 on the west flank by reason of this asymmetry is thus considerably more valuable in terms of petroleum potential than Section 34 on the east flank of the Cushing structure. The horizontal fluid interfaces in the deepest of the three sands, indicate that this unit is close to hydrostatic in nature and not a part of the same system as the two overlying hydrodynamic sands. With proper fluid and pressure data, these scenarios could be modeled using the versatile U, V, Z method.

There is always speculation on the rates at which tilting, displacement and flushing take place in response to hydrodynamic influences. The Cairo, Arkansas, field described by Goebel (1950) however is one example in which the timing of oil pool tilting can actually be determined. As illustrated in Figure 12.5, the reservoir occupies a domal structure. The original OWC is assumed, on the basis of the fresh oil staining pattern, to have been horizontal. Oil production and pumping of the same formation in the nearby Schuler pool appears to have created a water movement to the west. The creation of this artificial hydraulic gradient appears to have induced a 100-ft/mi oil–water contact tilt in the Cairo pool in a time period of 12 years, suggesting that pool tilting can occur astonishingly rapidly at least when produced by artificial systems.

The aspect of time is a major concern in the application of hydro-dynamics procedures in exploration and production-related problems. Little is known about just how rapidly or slowly the possible effects occur or how long they can be maintained. One speculates on whether the present locations of oil and gas pools reflect a response to today's hydro-logical regime or those that existed in the past. In other words, are today's pools where they are because of recent water movements or water move-

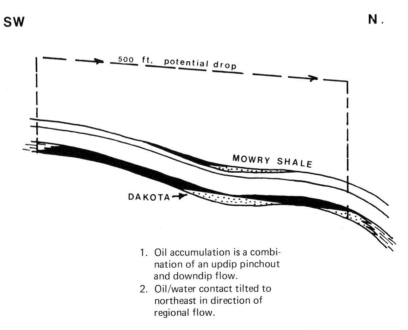

FIGURE 12.3. Cross section through the South Glenrock, Wyoming, oil field. (After Stone and Hoeger, 1973.)

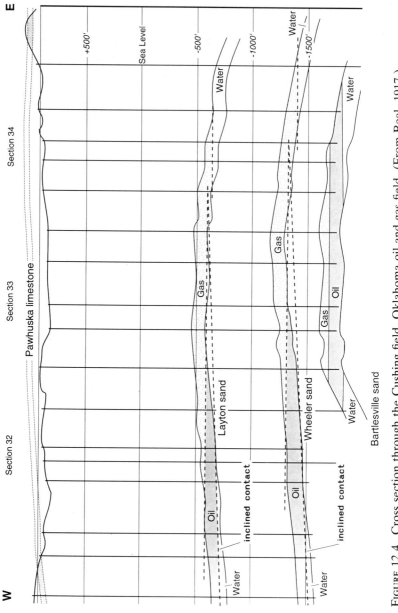

FIGURE 12.4. Cross section through the Cushing field, Oklahoma oil and gas field. (From Beal, 1917.)

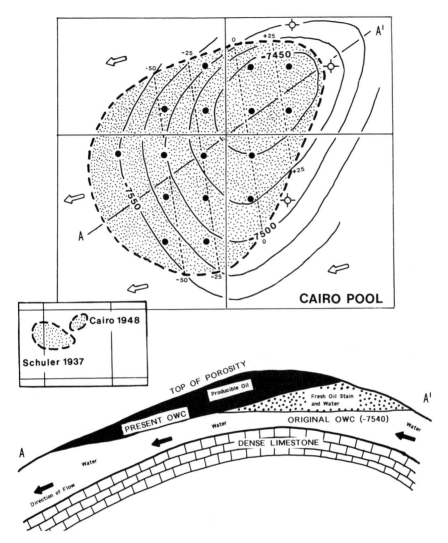

FIGURE 12.5. Map and cross section showing the Cairo Pool, Arkansas oil field. (Redrawn from Goebel, 1950.)

ments back in the Tertiary or Cretaceous? Also, presuming that the hydrogeology is always changing in response to tectonic and other geological influences, are hydrocarbon accumulations continually readjusting to fluctuating conditions. If so, how rapidly do significant positional changes occur? The Cairo example suggests that subsurface petroleum accumulations respond to changing water-flow conditions very quickly in terms of geologic time. Thus an understanding of present-day subsurface flow patterns should produce information relevant to current oil and gas pools.

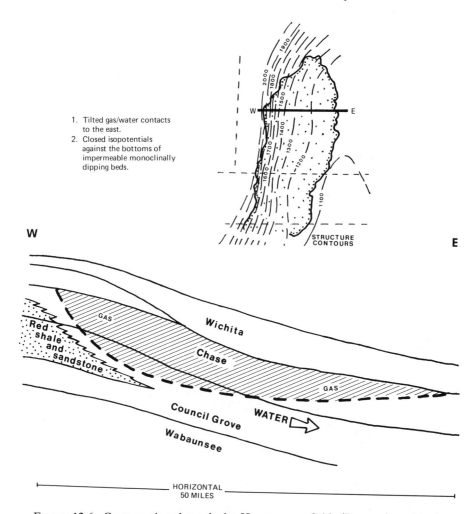

1. Tilted gas/water contacts to the east.
2. Closed isopotentials against the bottoms of impermeable monoclinally dipping beds.

FIGURE 12.6. Cross section through the Hugoton gas field. (Drawn from Pippin, 1968, and Hubbert, 1967.)

The Hugoton field in the midcontinental United States, shown in map and cross section in Figure 12.6, exhibits, according to Hubbert (1967) and Pippin (1968), a strongly tilted gas–water interface that dips to the east at a rate in excess of a 1,000 ft over a 60-mi distance. This tilt matches the well-documented west-to-east regional water flow pattern, strongly attesting to a hydrodynamic influence on North America's largest gas field. The Hugoton example strongly resembles the U, V, Z model shown in Figure 10.18, with downwarped gas isopotential surfaces located on the downflow side of a permeability barrier (in this case a series of interbedded red shales and sandstones) under conditions of downdip formation water flow. Once again verification with relevant data using the U, V, Z procedure would be very desirable.

Remarkably strong gravity-induced formation water movements created by the presence of surface recharge areas located at unusually high elevations are encountered in the mountainous regions of the western United States. As hydrogeological theory predicts, the hydrocarbon–water contacts in the associated petroleum reservoirs are often strongly tilted. One typical example is the Frannie oil field in the Big Horn Basin of Wyoming (Hubbert, 1953) illustrated in Figure 12.7. In this pool, the oil–water contact occurs at an elevation of 1,400 ft on the east margin of

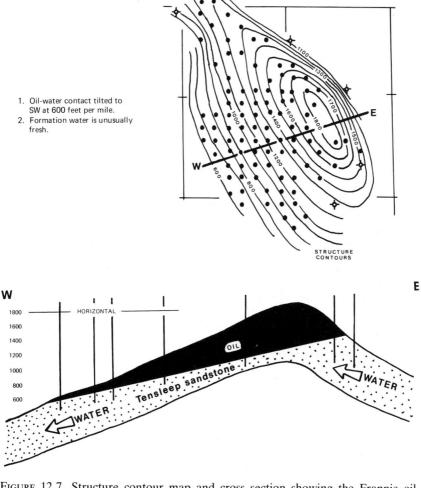

1. Oil-water contact tilted to SW at 600 feet per mile.
2. Formation water is unusually fresh.

FIGURE 12.7. Structure contour map and cross section showing the Frannie oil field in Wyoming. (Redrawn from Hubbert, 1953.)

the pool, while on the western edge its elevation is down close to 600 feet. This indicates a southwesterly drop of 800 ft in a mile-and-a-half—or about 600 ft/mi. Hubbert (1953) notes that the Tensleep formation waters flow in the same direction as that in which the field is tilted (SW) and that the magnitude of the dip on the OWC is very close to that which would be predicted from the hydrodynamic character of the water. An excellent treatment of potentiometric modeling in the Big Horn Basin and the effects of associated faults acting as either barriers or conduits is offered by Bredehoeft et al. (1992). For a regional picture of Big Horn Basin hydrogeology, see Figure 9.7.

The Weyburn oil field in Saskatchewan, Canada, provides a convincing example of the Hubbert oil-and-water density equation discussed in Chapter 9. The observed dip of the oil–water contact is very close to that predicted prior to the actual discovery of the pool. Hubbert (1967) notes that the oil–water contact in this producing field (defined as the 80% water saturation threshold) dips to the south at a rate of 40 to 60 ft/mi. The local direction of subsurface water flow is likewise to the south, with a potentiometric surface slope of about 10 ft/mi. The pool water and oil densities average 1.07 and 0.87 gm/cc respectively, which when used in the tilt amplification factor equation yields a TAF of 5.35. This results in an oil–water-contact dip value of 54 ft/mi, in agreement with the actual dip range above. The presence of the pool at a location where no obvious conventional trapping features are evident suggests that the unique occurrence of the Weyburn pool can be explained most convincingly largely as a product of hydrodynamic entrapment.

The Tin Fouye' pools in Algeria, a north-to-south trending cross section through which (from Chiarelli, 1978) is shown in Figure 12.8, occur in a series of northward plunging structural noses (see also Figures 9.10 and 9.11 after Chiarelli). The regional water flow is in a northerly direction from a mountainous recharge area to the south (Chiarelli, 1978) and the oil pools are strongly tilted in the same direction. The author pointedly observes and the cross section clearly demonstrates that without the northerly hydrodynamic gradient in this area, there would be relatively little potential for oil entrapment in these minimally closed features. The dip on the oil–water contact in these Ordovician pools is reported as being on the order of 10 m/km, which produces optimally large areas of closure between the tilted OWC and the reservoir roof.

A cross section through the famous Norman Wells oil pool, shown in Figure 12.9 (Hubbert, 1953) displays a tilted OWC, which is also shown contoured in Figure 9.3. The reservoir is essentially a fractured reef complex that has been tilted by tectonic events. The inclined interface between the Oil Saturated zone and the Unsaturated zone dips to the southwest at a rate of approximately 400 ft/mi. Whether this can be unequivocally related to hydrodynamic formation waters flowing in the same direction remains to be determined by examination of the local

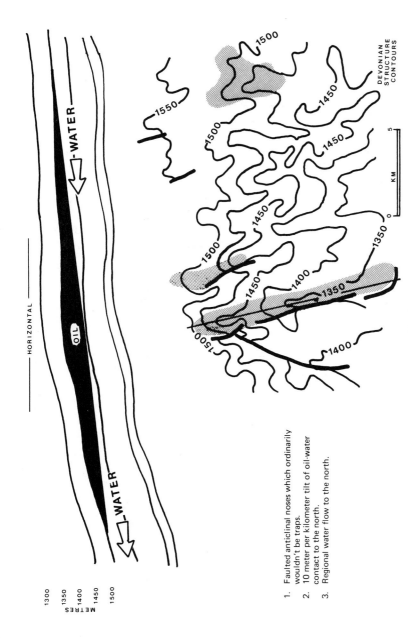

FIGURE 12.8. Cross section through and structure contour map of the Tin Fouye oil field of Algeria. (Redrawn from Chiarelli, 1978.)

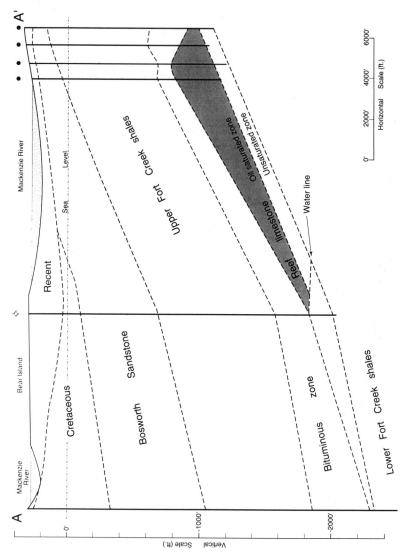

FIGURE 12.9. Cross section through the Norman Wells oil pool in the Northwest Territory of Canada. (From Hubbert, 1953.)

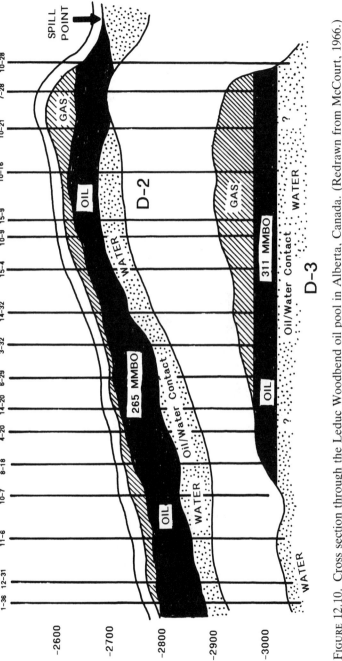

FIGURE 12.10. Cross section through the Leduc Woodbend oil pool in Alberta, Canada. (Redrawn from McCourt, 1966.)

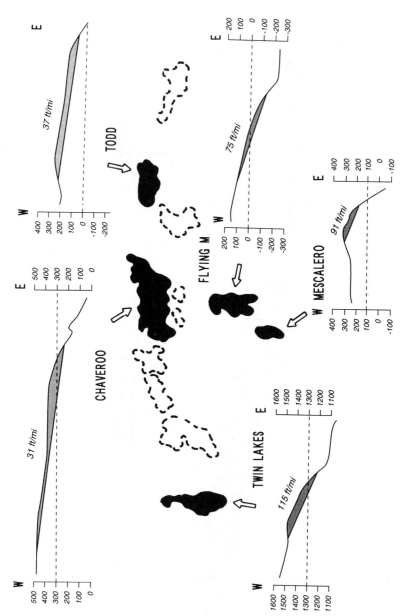

FIGURE 12.11. Cross sections showing dipping oil–water contacts in five oil pools in the San Andres area of New Mexico. (Redrawn from Keller, 1992.)

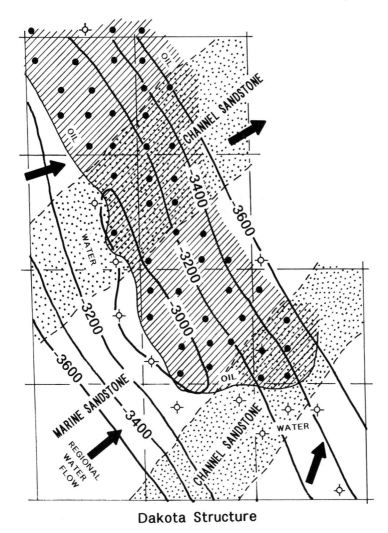

Dakota Structure

FIGURE 12.12. Structure contour and facies map of the hydrodynamically tilted and displaced South Cole Creek oil pool in Wyoming. (Redrawn from Moore, 1984.)

potentiometric relationships and U, V, Z modeling. But, as observed by Hubbert, the presence of the Franklin Mountain range as a potential water recharge area to the northeast of the pool location adds positive weight to the possible hydrodynamic character of this pool.

The oil–water contact in the Leduc Devonian Nisku pool in western Canada exhibits an undulating, but generally inclined, oil–water contact as illustrated in Figure 12.10. Whether the tilt can be attributed to hydrodynamic formation water has never been established. This and other cross

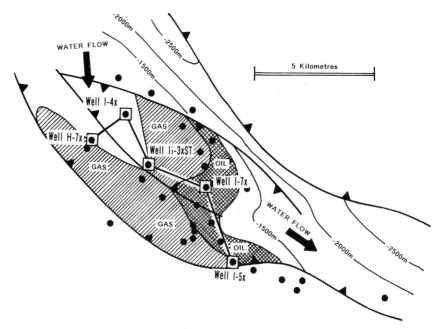

FIGURE 12.13. Map of the Iagifu/Hedinia field in Papua New Guinea showing the locations of the oil and gas legs. (Redrawn from Eisenberg, 1993.)

sections indicate that the OWC in the pool descends from −2500 ft down to approximately −2900 ft in a general northeast to southwest direction. An estimate of the maximum OWC dip is about 60 ft/mi.

In southeastern New Mexico, a series of tilted oil fields lying along a single producing trend have been described by Meissner (1974), Robichaud (1985) and most recently Keller (1992). Figure 12.11 shows the locations of these pools with east-west-oriented cross sections through them. The oil–water contacts all exhibit components of dip towards the east, which range from a high of 115 ft/mi to a low of 31 ft/mi. The first two authors have interpreted the tilted OWCs as evidence of hydrodynamic entrapment—or at least hydrodynamically-induced alteration of existing accumulations. Keller has interpreted these contacts as the result of porosity occlusion that essentially froze the oil in place, with subsequent tectonic uplift to the west creating the observed OWC dips to the east. Regional formation water flow is to the east as well, most likely as a result of the same event that created the elevated recharge area to the west. The observed dip magnitudes agree with those calculated using the Hubbert density equation, suggesting that the present pools are in equilibrium with the present flow regime. Keller also observes that there are no structural closures on any of the fields in the San Andres area and that structurally the accumulations are held in local "monoclinal flexures."

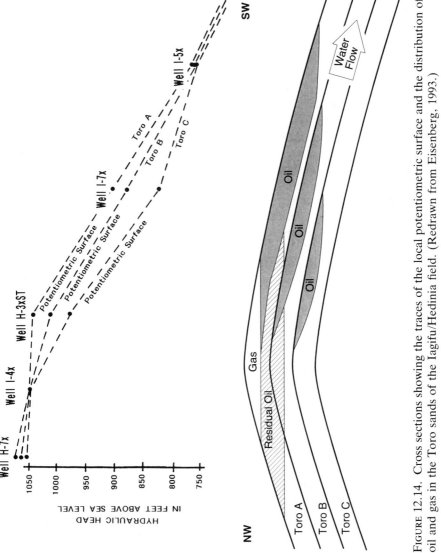

FIGURE 12.14. Cross sections showing the traces of the local potentiometric surface and the distribution of oil and gas in the Toro sands of the Iagifu/Hedinia field. (Redrawn from Eisenberg, 1993.)

A 440 ft/mi dipping oil–water-contact in the South Cole Creek field has been described by Moore (1984) as reflecting hydrodynamic oil entrapment in the Dakota sandstone. Figure 12.12 shows the oil accumulation offset to the downflow side (northeast) of the crest of the reservoir structure. Moore states and the map indicates that "the Dakota reservoir contains oil only on the east and north flanks of the structure, the south and west flanks are 'wet'." The regional water flow direction, inferred from a potentiometric map reinterpreted from Stone and Hoeger (1973), is essentially toward the northeast in the vicinity of the oil pool. In addition, the dip on the oil–water contact predicted from the Hubbert equation is 420 ft/mi (SGw = 1.05, SGo = .85) which matches (almost perfectly) the observed OWC dip. Two facies are indicated; a medium grain size channel sandstone that is highly productive and a fine grain size marine sandstone from which production is limited.

A very significant case illustrating downdip/downflow displacement of oil under gas is that of the Iagifu/Hedinia field described by Eisenberg (1993). This excellent, well-documented article, with its pressure-depth RFT gradient plots and potentiometric as well as OWC data, is essential reading for anyone interested in hydrodynamic oil entrapment and the relationships among structure, flow regime, and distributions of oil and gas. Figure 12.13 shows the major field outline and the thrust-faulted nature of the anticlinal structure. The author concludes that hydrodynamic flow has flushed the moveable oil from the northwest flank of the structure to the southeast, where it is currently held in the main block of the field as shown in cross section in Figure 12.14. Also shown in the cross section are traces of the potentiometric surfaces for each of the three reservoir sands, constructed from pressures recorded in the water legs of the appropriate wells. According to the author, the hydrodynamic character of the reservoir sands as inferred from the potentiometric gradients (Toro A, B, and C) has produced the tilted oil–water contacts, as well as displaced oil legs, as the theory dictates that it should.

As more and more explorationists become acquainted with the elements of hydrodynamic entrapment and their consequences on the size, shape, general nature, and configuration of oil and gas accumulations under these conditions, more and more previously unexplainable hydrocarbon occurrences become explainable. In the future, numerous new examples, such as those presented above, will find their way into the relevant literature and gain exposure to practicing exploration geologists and the petroleum industry. We are at present limited to the cases that appear in the public domain; it is intriguing to speculate on how many similarly appropriate examples of hydrodynamically affected entrapment are still "under wraps."

References

Following is a list of references on petroleum hydrogeology, hydrodynamics, formation pressure geology, and geopressures.

Adams, J.E. 1936. Oil pool of open reservoir type. *AAPG Bull.* **20,** pp. 780–796.

Allan, U.S. 1980. A model for the migration and entrapment of hydrocarbons. Abstract. AAPG Conf. on Seals for Hydrocarbons. Keystone, Colorado.

Arps, J.J. 1964. Engineering concepts useful in oil finding. *AAPG Bull.* **48,** no. 2, pp. 157–165.

Athy, L.F. 1930. Density, porosity, and compaction of sedimentary rocks. *AAPG Bull.* **14,** pp. 1–24.

Barker, Colin. 1972. Aquathermal pressuring—Role of temperature in development of abnormal-pressure areas. *AAPG Bull.* **56,** no. 10, pp. 2068–2071.

Barker, C., and B. Horsfield. 1973. Mechanical versus thermal cause of abnormally high pore pressures in shales. *AAPG Bull.* **66,** pp. 99–100.

Beal, C.H. 1917. Geologic structure of the Cushing Oil Field, Oklahoma. *USGS Bull.* **658,** p. 30.

Belitz, K., and J.D. Bredehoeft. 1988. Hydrodynamics of Denver Basin: Explanation of subnormal fluid pressures. *AAPG Bull.* **72,** pp. 1334–1359.

Berg, R.R. 1972. Oil column calculations in stratigraphic traps. *Gulf Coast Assoc. Geol. Trans.* **22,** pp. 63–66.

———. 1975. Capillary pressures in stratigraphic traps. *AAPG Bull.* **59,** pp. 939–956.

———. 1976. Trapping mechanisms for oil in Lower Cretaceous Muddy Sandstone at Recluse field, Wyoming. *Wyo. Geol. Assoc. Guidebook*, 29th Ann. Field Conf., pp. 261–272.

Berg, R.R., and A.R. Mitsdarffer. 1986. Effects of hydrodynamic flow on carbonate stratigraphic traps, Mission Canyon formation, Billings Nose fields, North Dakota. Abstract. *AAPG Bull.* **70,** no. 5, p. 564.

Berry, F.A.F. 1959. Hydrodynamics and geochemistry of the Jurassic and Cretaceous systems in the San Juan Basin. Unpubl. Ph.D. thesis, Stanford University.

———. 1973. High fluid potential in California Coast Ranges and their tectonic significance. *AAPG Bull.* **57,** no. 7, pp. 1219–1249.

Bethke, C.M. 1989. Modeling subsurface flow in sedimentary basins. *Geol. Rundschau* **78,** pp. 129–154.

Blair, E.S. et al., 1985. Potentiometric mapping from incomplete drill-stem test data; Palo Duro Basin Area, Texas and New Mexico. *Ground Water* **23,** no. 2, pp. 198–211.

Boatman, W.A. 1967. Measuring and using shale density to aid in drilling wells in high-pressure areas. *Jour. Petr. Tech.* **19**, no. 11, pp. 1423–1429.

Bogle, R.W., and W.B. Hansen. 1987. Knutson Field and its relationship to the Mission canyon oil play, south-central Williston Basin. Fifth Internat. Willist. Bas. Symp., *N.D. and Sask. Geol. Soc.*, pp. 242–252.

Bond, D.C. 1972. Hydrodynamics in deep aquifers of the Illinois basin. Circ. 470, Illinois State Geological Survey, Urbana, Illinois.

Bonham, L.C. 1980. Migration of hydrocarbons in a compacting basin. *In* W.H. Roberts and R.J. Cordell, eds., *Problems of petroleum migration*, pp. 69–88. AAPG Studies in Geology 10. Tulsa, Oklahoma: AAPG.

Borel, W.J., and R.L. Lewis. 1969. Ways to detect abnormal formation pressures. Part I *in* Geopressure predictions by log analysis. *Petrol. Eng.* **41**, no. 8, pp. 49–63.

Bowering, O.J.W. 1984. Hydrodynamics and hydrocarbon migration—A model for the Eromanga Basin. *APEA Jour.* **24**, pp. 227–236.

Bradley, J.S. 1975. Abnormal formation pressure. *AAPG Bull.* **59**, no. 6, pp. 957–973.

Bredehoeft, J.D. et al., 1963. Possible mechanisms for concentration of brines in subsurface formations. *AAPG Bull.* **47**, no. 2, pp. 5–10.

Bredehoeft, J.D., and R.R. Bennett. 1971. Potentiometric surface of the Tensleep sandstone in the Big Horn basin, west-central Wyoming. USGS map.

Bredehoeft, J.D., and B.B. Hanshaw. 1968. On the maintenance of anomalous fluid pressures, I and II. *Geolo. Soc. Amer. Bull.* **79**, pp. 1097–1106 and 1107–1122.

Bredehoeft, J.D., K. Belitz, and S. Sharp-Hansen. 1992. The hydrodynamics of the Big Horn Basin: A study of the role of faults. *AAPG Bull.* **76**, no. 4, pp. 530–546.

Bruce, C.H. 1973. Pressured shale and related sediment deformation; mechanism for development of regional contemporaneous faults. *AAPG Bull.* **57**, pp. 878–886.

Buhrig, C. 1989. Geopressured Jurassic reservoirs in the Viking graben: Modeling and geological significance. *Marine and Petroleum Geology* **6**, pp. 31–48.

Burst, J.F. 1969. Diagenesis of Gulf Coast clayey sediments and its possible relation to petroleum migration. *AAPG Bull.* **53**, no. 1, pp. 73–93.

Carlson, C.G., and J.E. Christopher (Eds.) 1987. Fifth International Williston Basin Symposium: Sask. Geol. Soc., Regina, **270**, pp. 25–30.

Carstens, H., and H. Dypvik. 1981. Abnormal formation pressure and shale porosity. *AAPG Bull.* **65**, pp. 346–350.

Chapman, R.E. 1980. Mechanical versus thermal cause of abnormally high pore pressures in shales. *AAPG Bull.* **64**, no. 12, pp. 2179–2183; (and reply) **66**, pp. 101–102.

———. 1981. Geology and water. The Hague: Martinus Nijhoff/Dr. W. Junk Publishers.

Chiarelli, A. 1978. Hydrodynamic framework of eastern Algerian Sahara—Influence on hydrocarbon occurrence. *AAPG Bull.* **62**, pp. 677–685.

Coll, D. 1991. Technology: The missing link. *Oilweek* **42**, no. 31, pp. 10–11.

Conybeare, C.E.B. 1965. Hydrocarbon-generation potential and hydrocarbon-yield capacity of sedimentary basins. *Bull. Can. Petro. Geol.* **13**, pp. 509–528.

————. 1970. Solubility and mobility of petroleum under hydrodynamic conditions, Surat Basin, Queensland. *Geol. Soc. Austr. Jour.* **16,** pt. 2, pp. 667–681.

Cousteau, H. 1977. Formation waters and hydrodynamics. *Jour. of Geochem. Expl.* **7,** pp. 213–241.

Cousteau, H., Chiarelli, A. et al., 1975. Classification hydrodynamique des bassins sedimentaries; Utilisation combinee avec d'autres methodes pour rationaliser l'exploration dans des bassins non productifs. *In* Proceedings, 9th World Petroleum Congress **2,** no. 4, pp. 105–119.

Curry, W.H., III. 1972. Resistivity mapping of sandstone stratigraphic traps. *Wyo. Geol. Assoc. Earth Sci. Bull.* **5,** no. 4, pp. 3–11.

Dahlberg, E.C. 1982. *Applied hydrodynamics in petroleum exploration.* New York: Springer-Verlag.

————. 1985. Formation pressure geology—Principles and practices. Unpubl. course manual. Calgary, Alberta: ECD Geological Specialists.

————. 1990. Grand banks hydrodynamics and formation pressure geology investigation: Pt. 2—Geopressure recognition and mapping. Private report.

————. 1992. Canadian offshore formation pressure geology: Geopressures and hydrodynamics of the Grand, Banks, Scotia Shelf and Beaufort Mackenzie Basin. Abstract. AAPG Conv. Calgary, Alberta.

————. 1993. One hydrodynamic oil prospect and one hydrodynamic oil play: Combinations of science, risk, geology and luck. Abstract. *Can. Soc. Petr. Geol. Bull.,* May 1993.

Daines, S.R. 1982. Aquathermal pressuring and geopressure evolution. *AAPG Bull.* **66,** no. 7, pp. 931–939.

Davidson, M.J. 1982. Toward a general theory of vertical migration. *Oil and Gas Jour,* June 21, p. 7.

De Mis, W.D. 1987. Hydrodynamic trapping in Mission Canyon reservoirs: Elkhorn Ranch field, North Dakota. *In* Ref. B, pp. 217–225.

Dickey, P.A. and W.C. Cox. 1977. Oil and gas reservoirs with subnormal pressures. *AAPG Bull.* **61,** pp. 2134–2142.

Dickey, P.A. et al., 1968. Abnormal pressures in deep wells in southwestern Louisiana. *Science* **160,** no. 3828, pp. 609–615.

Dickinson, G. 1953. Geological aspects of abnormal pressures in the Gulf Coast Region of Louisiana. *AAPG Bull.* **37,** pp. 410–432.

Doherty, M.G. et al., 1982. Methane production from geopressured aquifers. *Jour. Petr. Tech.* **34,** pp. 1591–1599.

Dorfman, M.H. 1982. The outlook for geopressured/geothermal energy and associated natural gas. *Jour. Petr. Tech.* **34,** pp. 1915–1919.

Downey, J.S. 1984. Geology and hydrology of the Madison Limestone and associated rocks in parts of Montana, Nebraska, North Dakota, South Dakota, and Wyoming. U.S. Geol. Surv. Prof. Paper 1273-G.

Downey, J.S., J.F. Busby, and G.A. Dinwiddie. 1987. Regional aquifers and petroleum in the Williston Basin region of the United States. *In* Ref. C, pp. 299–313.

Downey, M.W. 1984. Evaluating seals for hydrocarbon accumulations. *AAPG Bull.* **68,** no. 11, pp. 1752–1763.

Edie, R.W. 1955. The inclined oil-water contact at the Joarcam Field. *Jour. Alberta Soc. of Petrol. Geol.* **3,** no. 7, pp. 99–103.

Eisenberg, L.I. 1993. Hydrodynamic character of the Toro sandstone, Iagifu/ Hedinia area, Southern Highlands Province, Papua New Guinea. Proc. Second PNG Petroleum Convention, Port Moresby, pp. 447–458.

Eremenko, N.A., and I.M. Michailov. 1974. Hydrodynamic pools at faults. *Bull. Gan. Pet. Geol.* **22**, no. 2, pp. 106–118.

Ettinger, M.I., and Y. Langozky. 1969. Hydrodynamics of the Mesozoic formations in the Northern Negev, Israel. *Geol. Surv. Isr. Bull.* **46**, pp. 1–32.

Fanshawe, J.R. 1960. Significance of interruptions to hydrodynamics in northern rocky Mountain province, U.S.A. Internatl. Geol. Congr., 21st Sess., Pt. 11, Copenhagen, pp. 7–18.

Fatt, I. 1958. Compressibility of sandstones at low to moderate pressures. *AAPG Bull.* **42**, pp. 1924–1957.

Fayers, F.S., and J.W. Sheldon. 1962. The use of high-speed digital computers in the study of the hydrodynamics of geologic basins. *Jour. Geophys. Res.* **67**, no. 6, pp. 2421–2431.

Fertl, W.H. 1976. Abnormal formation pressures. *Developments in petroleum science 2.* New York: Elsevier.

Fertl, W.H., and D.J. Timko. 1972. How downhole temperatures, pressures affect drilling. *World Oil*, June, pp. 67–70.

Foster, J.B., and H.E. Whalen. 1966. Estimation of formation pressures from electrical surveys, offshore Louisiana. *Jour. Petrol. Tech.*, pp. 165–171.

Foster, W.R., and H.C. Custard. 1980. Smectite-illite transformation—Role in generating and maintaining geopressure. *AAPG Bull.* **64**, p. 708.

Fowler, W.A. 1970. Pressures, hydrocarbon accumulation, and salinities— Chocolate Bayou Field, Brazoria County, Texas. *Jour. Petrol. Tech.*, April, pp. 411–423.

Frederick, W.S. 1967. Planning a must in abnormally pressured areas. *World Oil* **164**, no. 4, pp. 73–78.

Freeze, G.I., and W.S. Frederick. 1967. Abnormal drilling problems in the Anadarko Basin. Preprint, Div. of Production, Am. Petrol. Inst., Paper 851-41-6.

Frenzel, H.N., and E.R. Sharp. 1975. The Indian Basin Upper Pennsylvanian gas field, Eddy County, New Mexico. Trans. SW Sect AAPG Mtg., El Paso Texas Geol. Soc., pp. 149–167.

Garven, G. 1985. The role of regional fluid flow in the genesis of the Pine Point deposit, Western Canada sedimentary Basin. *Econ. Geol.* **80**, no. 2, pp. 307–324.

Geis, R.M. 1984. Case history for a major Alberta Deep Basin gas trap; the Cadomin Formation. *AAPG Memoir* **38**, pp. 115–140.

Gill, D. 1979. Differential entrapment of oil and gas in Niagaran pinnacle—Reef belt of northern Michigan. *AAPG Bull.* **63**, no. 4, pp. 608–620.

Goebel, L.A. 1950. Cairo Field, Union County, Arkansas. *AAPG Bull.* **34**, no. 10, pp. 1954–1980.

Goolsby, S.M., L. Druyff, and M.S. Fryt. 1988. Trapping mechanisms and petrophysical properties of the Permian Kaibab formation, South-Central Utah. RMAG Carbonate Symp., Denver, Colorado, pp. 193–211.

Gorrell, H.A. 1963. Some indications of subsurface features from ground water data. Unpubl.

Grauten, W.F. 1965. Fluid relationships in Delaware Mountain Sandstone. *AAPG Bull.* pp. 294–307.

Gretener, P.E. 1969. Fluid pressure in porous media—Its importance in geology; a review. *Bull. Canad. Petr. Geol.* **17**, pp. 255–295.

———. 1978. Pore pressure: Fundamentals, general ramifications and implications for structural geology. AAPG Course Note Series 4.

Griffin, D.G., and D.A. Bazer. 1968. A comparison of methods for calculating pore pressures and fracture gradients from shale density measurements using the computer. Paper 2166, SEPM, Houston.

Habermann, Ben. 1960. A study of the capillary pressure-hydrodynamic relationship of oil accumulation in stratigraphic traps. *CIMM Bull.*, October, pp. 811–817.

Ham, H.H. 1966. New charts help estimate formation pressures. *O & G Jour.* **64**, no. 51, pp. 58–63.

Hannon, N.M. 1987. Subsurface waterflow patterns in the Canadian portion of the Willistion Basin. *In* Ref. C, pp. 313–321.

Hannon, N.M. 1967. Regional aquifer ties Rainbow pools. *Canadian Petroleum*, September, pp. 32–34.

Hannon, J.S., and R. Sassen. 1988. Deep basin hydrodynamics of the Louisiana Gulf Coast: Implications for oil and gas migration. Abstract. Ninth Ann. Res. Conf, Gulf Coast Sect., SEPM Found., Abstracts booklet.

Hanshaw, B.B., and J.D. Bredehoeft. 1968. On the maintenance of anomalous fluid pressures; II. Source layer at depth. *GSA Bull.* **79**, pp. 1107–1122.

Harbaugh, J.W. 1970. Trend-surface mapping of hydrodynamic oil traps with the IBM 7090/94 computer. *Quarterly Colo. Sch. Mines*, pp. 557–578.

Harkins, K.L., and J.W. Baugher. 1969. Geological significance of abnormal formation pressures. *Jour. Petr. Tech.* **21**, pp. 961–966.

Haun, J.D. 1960. Potentiometric surface map, Phosphoria and Tensleep Formations, southwest Big Horn Basin, Wyoming. Unpubl. map.

———. 1970. Overpressures in the eastern Wind River Basin, Wyoming. Unpubl. report.

———. 1982. Pressure analysis and hydrodynamics in petroleum exploration. Australian Mineral Foundation, Workshop Course 183/82.

Hedberg, H.D. 1926. The effect of gravitational compaction on the structure of sedimentary rocks. *AAPG Bull.* **10**, pp. 1035–1072.

———. 1974. Relation of methane generation to undercompacted shales, shale diapirs, and mud volcanoes. *AAPG Bull.* **58**, no. 4, pp. 661–673.

———. 1980. Methane generation and petroleum migration. *In* W.H. Roberts and R.J. Cordell, eds., *Problems of petroleum migration*, pp. 179–206. AAPG Studies in Geology No. 10: Tulsa, Oklahoma, AAPG.

Hemphill, C.R., R.I. Smith, and F. Szabo. 1970. Beaverhill Lake reefs, Swan Hills Area, Alberta. *AAPG Memoir.* **14**, Tulsa, Okla. pp. 50–90.

Herring, E.A. 1973. North Sea abnormal pressures determined from logs. *Petrol. Eng.* **45**, pp. 72–84.

Hill, G.A., W.A. Colburn, and J.W. Knight. 1961. Reducing oil-finding costs by use of hydrodynamic evaluations. *In Economics of petroleum exploration*, pp. 38–69. Englewood Cliffs, N.J.: Prentice-Hall.

Hitchon, B. 1969. Fluid flow in the Western Canada sedimentary basin. 1. Effect of topography. *Water Resources Research* **5**, no. 1, pp. 186–195. 2. Effect of geology. *Water Resources Research* **5**, no. 2, pp. 460–469.

————. 1984. Geothermal gradients, hydrodynamics and hydrocarbon occurrences, Alberta, Canada. *AAPG Bull.* **68,** pp. 713–743.

Hitchon, B., and J. Hays. 1971. Hydrodynamics and hydrocarbon occurrences, Surat Basin, Queensland, Australia. *Water Resour. Res.* **7,** no. 3, pp. 658–676.

————. 1975. Hydrodynamics and geochemistry, Surat Basin and underlying sediments, Queensland, Australia. Unpubl.

Horne, R.H. et al., 1986. The use of microcomputers in well test data acquisition and analysis. SPE Symp. on Petroleum Industry Applications of Microcomputers, Colorado, June.

Horner, D.R. 1951. Pressure build-up in wells. Proceedings of the Third World Petroleum Cong., Sec. II, pp. 503–521.

Hottman, C.E., and R.K. Johnson. 1965. Estimation of formation pressures from log-derived shale properties. *Jour. Petrol. Tech.*, June, pp. 717–722.

Hubbert, M.K. 1940. The theory of ground-water motion. *Jour. Geol.* **48,** no. 8, pp. 785–944.

————. 1967. Application of hydrodynamics to oil exploration. Paper RPCR-4, 7th World Petrol. Congr., Mexico City.

————. 1953. Entrapment of petroleum under hydrodynamic conditions. *AAPG Bull.* **37,** no. 8, pp. 1954–2026.

Hugo, K.J. 1990. Mechanisms of groundwater flow and oil migration associated with Leduc reefs. *Bull. Canad. Petr. Geol.* **38,** pp. 307–319.

Hunt, J.M. 1990. Generation and migration of petroleum from abnormally pressured fluid compartments. *AAPG Bull.* **74,** no. 1, pp. 1–12.

Illing, V.C. 1938. The origin of pressure in oil pools. *In The science of petroleum*, v. 1, pp. 224–229. London: Oxford University Press.

Issar, A., and E. Rosenthal. 1968. The artesian basins of the central plateau of Iran. *In* Proc. Internat. Assoc. Hydrol., 23rd Sess. Internat. Geol. Cong., Geol. Surv. of Isreal, Jerusalem.

Jacquin, C., and M. Poulet. 1970. Study of the hydrodynamic pattern in a sedimentary basin subject to subsidence. SPE of AIMS Paper SPE2988.

————. 1973. Essai de restitution des conditions hydrodynamiques regnant dans un bassin sedimentaire au cours de son evolution. *Revue de l'Institut Francais du Petrole* **28,** no. 3, pp. 269–297.

Jones, P.H. 1969. Hydrodynamics of geopressures in the northern Gulf of Mexico Basin. *Jour. Petrol. Tech.*, July, pp. 803–810.

Jorden, J.R., and O.J. Shirley. 1966. Application of drilling performance data to overpressure detection. *Jour. Petr. Tech.* **18,** pp. 1387–1394.

Kartsev, A.A., and S.B. Vagin. 1962. Paleohydrogeological studies of the origin and dissipation of oil and gas accumulations in the instance of Cis-Caucasian Mesozoic deposits. *Internat. Geol. Rev.* **6,** no. 4, pp. 644–655.

Keller, D.R. 1992. Evaporite geometries and diagenetic traps, Lower San Andres, Northwest Shelf, New Mexico. *Trans. SW Sect. AAPG Conv.*, Midland, Texas. pp. 183–193.

Kincheloe, R.L., and J. Scott. 1974. What Lone Star found at 31,441 feet. *Petr. Eng. Int.* **46,** pp. 29–32.

Kok, P.C., and Thomeer. 1955. Abnormal pressures in oil and gas reservoirs. *Geol. en Mijn.* **17,** pp. 207–216.

Kortsenshteyn, V.N. 1964. Predicting oil and gas by means of groundwater investigations at depth and estimating the potential reserves of oil and gas.

Doklady Akademii Nauk SSSR, Earth Science Section **58,** no. 4, pp. 856–859.

Krayushkin, V.A. 1967. Effect of capillary pressure on distribution of gas and oil occurrences in lateral migration of oil and gas. *Int. Geol. Rev.* **10,** no. 8, pp. 10–15.

Kudryakow, V.A. 1983. Piezometric minima and their role in the formation and distribution of hydrocarbon accumulations. *Doklady Akademii Nauk SSSR* **207,** no. 6, pp. 1424–1426.

Larberg, G.M. 1976. Computer analysis of drill stem test data. Unpubl. manuscript, Texas A & M University.

———. 1981. Downward vertical fluid flow in subsurface: Implications at Kitty Field, Powder River Basin, Wyoming. *AAPG Bull.* **65,** no. 1, pp. 1869–1887.

Larson, T.G. 1971. Hydrodynamic interpretation of mid-continent. *In* F. J. Adler et al., Future petroleum provinces of the mid-continent, Region 7. *AAPG Memoir 15*, v. 2, pp. 1043–1046.

Lewis, C.R., and S.C. Rose. 1970. A theory relating high temperatures and overpressures. *Jour. Petrol. Tech.*, January.

Lin, J.T.C. 1981. Hydrodynamic flow in lower cretaceous Muddy sandstone, Gas Draw field, Powder River Basin, Wyoming. *The Mountain Geologist* **18,** no. 4, pp. 78–87.

Lucas, P.T., and J.M. Drexler. 1975. Altamont-Bluepell: A major fractured and overpressured stratigraphic trap, Uinta Basin, Utah. RMAG Symp. on Deep Drilling Frontiers in the Central Rocky Mountains, pp. 265–273.

Magara, K. 1975a. Reevaluation of Montmorillonite dehydration as cause of abnormal pressure and hydrocarbon migration. *AAPG Bull.* **59,** no. 2.

———. 1975b. Importance of aquathermal pressuring effect in Gulf Coast. *AAPG Bull.* **59,** no. 10.

———. 1975c. Importance of hydrodynamic factor in formation of Lower Cretaceous traps, Big Muddy-South Glenrock area, Wyoming; discussion. *AAPG Bull.* **59,** pp. 890–893.

———. 1978. Compaction and fluid migration: Practical petroleum geology. Amsterdam: Elsevier.

Martin, B. 1967. Variation des cagacteres chimiques des eaux dans les niveaux poreux du Bassin de Parentis. *Chron. Hydrogeol.* **12,** pp. 77–89.

Masroua, L.F. 1973. Patterns of pressure in the Morrow sands of central Oklahoma. Master's thesis, University of Tulsa.

Masters, J.A. 1991. Deep basin gas trap, western Canada. *AAPG Bull.* **63,** no. 2, pp. 152–181.

Matthews, W.R., and J. Kelly. 1967. How to predict formation pressure and fracture gradient from electric and sonic logs. *Oil and Gas Jour.* **65,** no. 8, pp. 92–106.

McConnell, C.L. 1985. Salinity and temperature anomalies over structural oil fields, Carter County, Oklahoma. *AAPG Bull.* **69,** no. 5, pp. 781–787.

McNeal, R.P. 1961. Hydrodynamic entrapment of oil and gas in Bisti field, San Juan County, New Mexico. *AAPG Bull.* **45,** pp. 315–326.

———. 1965. Hydrodynamics of the Permian basin. *In* Fluids in subsurface environments. *AAPG Memoir* **4,** pp. 308–326.

McPeek, L.A. 1981. Eastern Green River Basin: A developing giant gas supply from deep, overpressured Upper Cretaceous sandstones. *AAPG Bull.* **65,** no. 6, pp. 1078–1098.

Meinhold, R. 1971. Hydrodynamic control of oil and gas accumulation as indicated by geothermal, geochemical and hydrological distribution patterns. *Proc. 8th World Petrol. Cong.* **2,** pp. 55–66.

Meissner, F.F. 1974. Hydrocarbon accumulation in San Andres Formation of Permian Basin, southeast New Mexico and West Texas (abstract). *AAPG Bull.* **58,** pp. 909–910.

———. 1980. Examples of abnormal pressure produced by hydrocarbon generation. *AAPG Bull.* **64,** p. 749.

———. 1981. Abnormal pressures produced by hydrocarbon generation and maturation and their relation to processes of migration and accumulation. Abstract. *AAPG Bull.* **65,** no. 11, p. 2467.

Meyers, J.D. 1968. Differential pressures, a trapping mechanism in Gulf Coast oil and gas fields. *Gulf Coast Assoc. Geol. Socs. Trans.* **18,** pp. 56–80.

Michaels, A.S., and E.A. Hauser. 1950. Interfacial properties of hydrocarbon-water systems. *Jour. Phys. and Colloidal Chemistry* **55,** pp. 408–421.

Mitsdarffer, A.R. 1985. Hydrodynamics of Mission Canyon Formation in the Billings Nose Area, North Dakota. Master's thesis, Texas A & M University.

Moore, R.W. 1984. Hydrodynamic control on oil entrapment in channel sandstones of the Dakota sandstone, South Cole Creek Field, Converse County, Wyoming. *The Mountain Geologist.* **21,** no. 3, pp. 105–113.

Munn, M.J. 1909. The anticlinal and hydraulic theories of oil and gas accumulation. *Econ. Geol.* **4,** no. 6, pp. 509–529.

Murphy, W.C. 1971. The interpretation and calculation of formation characteristics from formation test data. Duncan, Oklahoma: Halliburton Services.

Murray, G.H. 1959. Examples of hydrodynamics in the Williston Basin at Poplar and North Tioga Fields. *AAPG Geol. Record,* Rocky Mtn. Sect., pp. 55–60.

Muskat, M. 1932. Problems of underground water-flow in the oil industry. *Am. Geophys. Un. Trans.,* pp. 399–401.

Myer, J.D. 1968. Differential pressures: A trapping mechanism in Gulf Coast oil and gas fields. *Gulf Coast Assoc. Geol. Socs. Trans.* **18,** pp. 56–80.

Neglia, S. 1979. Migration of fluids in sedimentary basins. *AAPG Bull.* **63,** pp. 573–597.

Neuzil, C.E., and D.W. Pollock. 1983. Erosional unloading and fluid pressures in hydraulically "tight" rocks. *Jour. Geol.* **91,** no. 2, pp. 179–193.

Olsen, H.W. 1972. Liquid movement through kaolinite under hydraulic, electric and osmotic gradients. *AAPG Bull.* **56,** no. 10.

Pedry, J.J. 1975. Tensleep sandstone stratigraphic-hydrodynamic traps, northest Bighorn Basin, Wyoming. 27th Ann. Field Conf., *Wyo. Geol. Assoc. guidebook,* pp. 117–127.

Pelissier, J. et al., 1980. Study of hydrodynamic activity in the Mishrif fields of offshore Iran. *Jour. Petrol. Tech.,* SPE paper 7508, pp. 1043–1052.

Perry, E.A., and J. Hower. 1972. Late-stage dehydration in deeply buried pelitic sediments. *AAPG Bull.* **56,** pp. 2013–2021.

Peterson, J.A. et al. (Eds.) 1987. Williston Basin: Anatomy of a cratonic oil province. Denver: RMAG.

Petroleum Research. 1960. Geological significance of pressure gradients, capillarity, and relative permeability. Research report A-O. Unpubl.

Pippin. L. 1968. Panhandle-Hugoton Field, Texas-Oklahoma-Kansas. *AAPG Memoir.* **14,** pp. 204–222.

————. 1970. Panhandle-Hugoton field, Texas-Oklahoma-Kansas—The first fifty years. *AAPG Memoir 14*, Geology of Giant Petroleum Fields, pp. 204–222.

Pirson, S.J. 1967. How to use well logs to seek hydrodynamically trapped oil. *World Oil* **164**, no. 5, pp. 100–106.

Plumley, W.J. 1980. Abnormally high fluid pressure: Survey of some basic principles. *AAPG Bull.* **64**, pp. 414–422.

Powers, M.C. 1967. Fluid-release mechanisms in compacting marine mudrocks and their importance in oil exploration. *AAPG Bull.* **51**, no. 7, pp. 1240–1254.

Powley, D.E. 1990. Pressures and hydrogeology in petroleum basins. *Earth Science Reviews* **29**, pp. 215–226.

Prendergast, R.D. 1965. Correlating stratigraphy with the use of drill stem test pressures. Lynes United Services.

————. 1969. Correlating Cardium stratigraphy using static bottom hole pressures. *Canad. Well Log. Soc. Jour.* **2**, no. 1, pp. 19–27.

Price, L.C. 1976. Aqueous solubility of petroleum as applied to its origin and primary migration. *AAPG Bull.* **60**, pp. 213–244.

Putnam, P.E., and T.A. Oliver. 1980. Stratigraphic traps in channel sandstones in the Upper Mannville (Albian) of East-Central Alberta. *CSPG Bull.* **28**, no. 4, pp. 489–508.

Rehm, R., and R. McClendon. 1971. Measurement of formation pressure from drilling data. SPE 3601.

Reynolds, E.B. 1970. Predicting overpressured zones with seismic data. *World Oil* **171**.

Rich, J.L. 1921. Moving underground water as a primary cause of the migration and accumulation of oil and gas. *Econ. Geol.* **16**, no. 6, pp. 347–371.

Roach, J.W. 1963. How to determine oil-water contacts using one point control. *World Oil*, December, pp. 87–90.

————. 1965a. How to apply fluid mechanics to petroleum exploration, Part 1. *World Oil* **160**, pp. 71–75.

————. 1965b. How to apply fluid mechanics to petroleum exploration, Part 2. *World Oil* **160**, pp. 131–134.

Roberdeau, S.W. 1962. An electrical-analog model and its application to the study of the effect of hydrodynamics on the entrapment of petroleum. Unpubl. Master's thesis, Colorado School of Mines.

Robichaud, S. 1985. San Andres Formation from the Flying M Field in Lea County, N.M. In *Permian Basin Cores: A Workshop*. SEPM Workshop 3, pp. 110–121.

Rogers, L. 1966. Shale-density logs help detect overpressure. *O & G Jour.* **64**, no. 37, pp. 126–130.

Russell, W.L. 1956. Tilted fluid contacts in the mid-continent region. *AAPG Bull.* **40**, no. 11, pp. 2644–2688.

————. 1961. Reservoir water resistivities and possible hydrodynamic flow in Denver Basin. *AAPG Bull.* **45**, no. 12, pp. 1925–1940.

————. 1972. Pressure-depth relations in Appalachian region. *AAPG Bull.* **56**, pp. 528–536.

Sahay, B. 1972. Abnormal subsurface pressures, their origin and methods employed for prediction in India. SPE 3900, 3rd Symp. on Abnormal Subsurface Pore Pressure, Louisiana State University, pp. 153–159.

Saidi, A.M., and L.E. Martin. 1965. Applications of reservoir engineering in the development of Iranian reservoirs. A.C.A.F.E. Symp. on Petroleum, Tokyo, Japan.

Schowalter, T.T. 1979. Mechanics of secondary hydrocarbon migration and entrapment. *AAPG Bull.* **63**, no. 5, pp. 723–760.

Schwab, R. 1965. Logging important aspect of hydrodynamic studies. *Oilweek* **16**, no. 25, pp. 36–37.

Sharp, G.C. 1976. Reservoir variations at Upper Valley Field, Garfield County, Utah. RMAG, 1976 Symp., pp. 325–343.

Smith, D.A. 1966. Theoretical considerations of sealing and non-sealing faults. *AAPG Bull.* **50**, no. 2, pp. 363–374.

Spencer, C.W. 1987. Hydrocarbon generation as a mechanism for overpressuring in the Rocky mountain region. *AAPG Bull.* **71**, pp. 207–327.

Stone, D.S., and R.L. Hoeger. 1973. Importance of hydrodynamic factor in formation of Lower Cretaceous combination traps, Big Muddy-South Glenrock area, Wyoming. *AAPG Bull.* **57**, no. 9, pp. 1714–1733.

Terzaghi, K. 1936. Simple tests determine hydrostatic uplifz. *Eng. New Record* **116**, 872–875.

Thomeer, J.G.M.A., and J.A. Bottema. 1961. Increasing occurrence of abnormally high reservoir pressures in boreholes and drilling problems resulting therefrom. *AAPG Bull.* **45**, pp. 1721–1730.

Timko, K.J., and W.H. Fertl. 1971. Relationship between hydrocarbon accumulation and geopressure and its economic significance. *Jour. Petr. Tech.* **23**, pp. 923–933.

Toth, J. 1987. Petroleum hydrogeology: A new basic in exploration. *World Oil*, September.

Toth, J., and T. Corbet. 1983. Investigations of the relations between formation fluid dynamics and petroleum occurrences, Taber area, Alberta. Alberta: Alberta ENR Edmonton.

Toth, J., and K. Rakhit. 1988. Exploration for reservoir quality rock bodies by mapping and simulation of potentiometric surface anomalies. *Cdn. Soc. Petr. Geol. Bull.* **36**, n. 4, pp. 362–378.

van Everdingen, R.O. 1968. Studies of formation waters in Western Canada: Geochemistry and hydrodynamics. *Can. Jour. of Earth Sci.* **5**, pp. 523–543.

van Poollen, H.K., and S.J. Bateman. 1958. Application of DST to hydrodynamic studies. *World Oil*, July.

Varley, C.J. 1984. The Cadomin formation: A model for the deep basin type gas trapping, mechanism. *Can. Soc. Petrol. Geol. Memoir* **9**, pp. 471–484.

Vugrinovich, R. 1988. Relationships between regional hydrogeology and hydrocarbon occurrences in Michigan, USA. *Jour. Petrol. Geol.* **11**, pp. 429–442.

Wallace, W.E. 1962. Water production from abnormally pressured gas reservoirs in south Louisiana. *Trans Gulf Coast Assn. Geol. Soc.* **12**, pp. 187–193.

———. 1964. Will induction log yield pressure data? *Oil and Gas Jour.* **62**, no. 37, pp. 124–126.

Wells, P.R.A. 1988. Hydrodynamic trapping in the Cretaceous Nahr Umr Lower Sand of the North area, offshore Qatar. *Jour. Petrol. Tech.* **40**, pp. 357–362.

Willis, D.G. 1961. Entrapment of petroleum. *In* G.B. Moody, ed., *Petroleum exploration handbook*, pp. 1–68. New York: McGraw-Hill.

Young, A., and P.F. Low. 1965. Osmosis in argillaceous rocks. *AAPG Bull.* **49,** no. 7, pp. 1005–1007.

Yuster, S.Y. 1953. Some theoretical considerations of tilted water tables. *A.I.M.E. Petrol. Trans.* **198,** pp. 149–156.

Zawisza, L. 1986. Hydrodynamic conditions of hydrocarbon accumulation exemplified by the Carboniferous formation in the Lublin synclinorium, Poland. *Soc. Petrol. Eng.,* Formation Evaluation, pp. 286–294.

List of Symbols and Abbreviations Appearing in the Text

API	American Petroleum Institute
B	buoyancy force
D	density or depth
D_G	density of gas
D_H	density of hydrocarbon
D_o	density of oil
D_w	density of water
dP/dZ	pressure gradient; rate of change of pressure relative to rate of change of vertical elevation
DST	drill stem test
ΔP	increment of change in pressure
$\Delta \Phi$	increment of change in potential energy
ΔX	increment of change in distance
ΔZ	increment of change in elevation
E_G	net force concentrated at a point in gas
E_o	net force concentrated at a point in oil
E_w	net force concentrated at a point in water
F_w	directional flow component
FP	flow pressure (DST)
FSIP	final shut-in pressure (DST)
G	gas
g	acceleration of gravity
GWC, G/W	gas–water contact
GCXW	gas-cut salt water
grad P	pressure gradient
grad Φ	potential energy gradient
H or h	height or "hydraulic head"
h-Z	height or hydraulic head relative to a specific measurement
H_o or h_o	hydraulic head for oil
HP	hydrostatic (mud) pressure (DST)
H_w or h_w	hydraulic head for water

258

ISIP	initial shut-in pressure (DST)
KB	Kelly Bushing; reference elevation on a rig floor
kPa	kilopascal
m	mass
O	oil
OWC, O/W	oil–water contact
OCXW	oil-cut salt water
P	pressure
p	pressure
P-D	pressure–depth
P_c	capillary pressure
ppm	parts per million
psi	pounds per square inch
Φ	potential energy
R	resultant force
RD	recorder depth
SG	specific gravity
SIP	shut-in pressure (DST)
SL	sea level
U	potential energy of hydrocarbon at a point
U_G	potential energy of gas at a point
U_o	potential energy of oil at a point
V	potential energy of water at a point
v	volume
V_G	potential energy of water relative to gas at a point
V_o	potential energy of water relative to oil at a point
w	water
w	weight
XW	salt water
Z	elevation

Appendix A

Answers to Problems

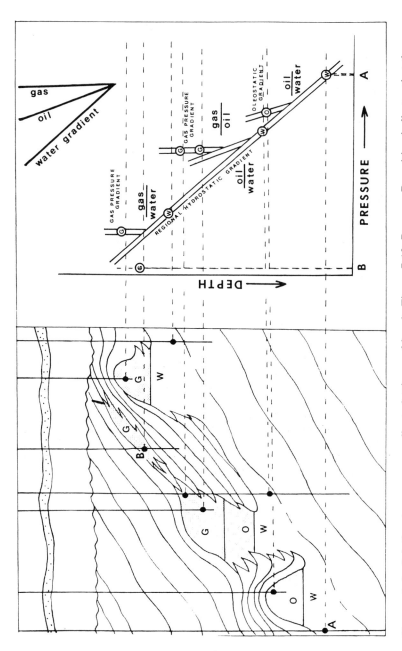

FIGURE A.1. Solution-to-pressure gradient plotting problem in Figure 7.10. Pressure B position indicates that the gas reservoir is not a part of the main system, that is, a younger reservoir.

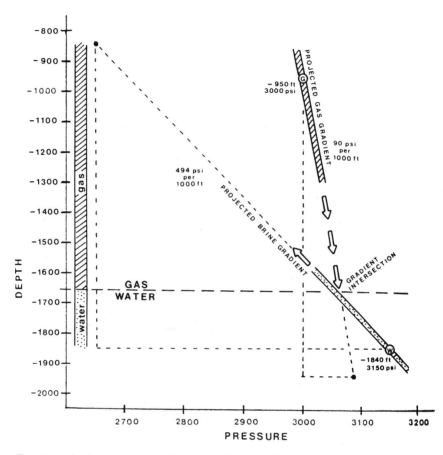

FIGURE A.2. Solution to the first part of the problem in Figures 7.11 and 7.12. The brine gradient is projected upward from the water test point and the gas gradient downward from the estimated crest point. The intersection of the two gradients defines the likely gas–water transition zone, which can be plotted onto the structure map for evaluation purposes.

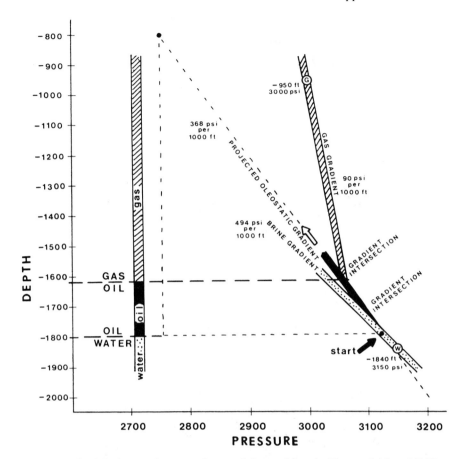

FIGURE A.3. Solution to the second part of the problem in Figures 7.11 and 7.12. The oleostatic (oil pressure) gradient is projected upward from the assumed oil–water contact at 1,800 ft, using proper slope. The gas gradient is projected downward to the point of intersection, which corresponds to the elevation of the gas–oil interface. Elevations of gas–oil and oil–water contacts are then mapped.

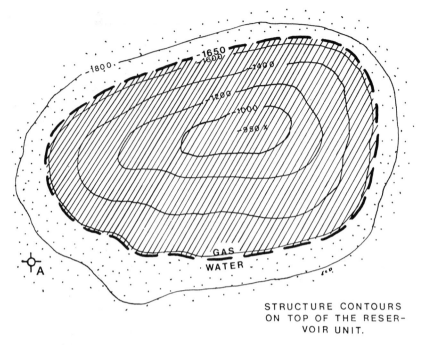

STRUCTURE CONTOURS
ON TOP OF THE RESER-
VOIR UNIT.

FIGURE A.4. Gas–water contact mapped onto the seismic structure of Figure 7.11.

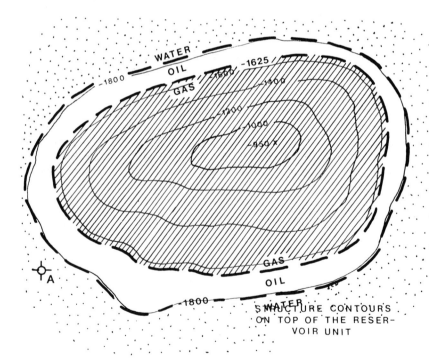

STRUCTURE CONTOURS
ON TOP OF THE RESER-
VOIR UNIT

FIGURE A.5. Gas–oil and oil–water contacts mapped onto the seismic structure of Figure 7.11.

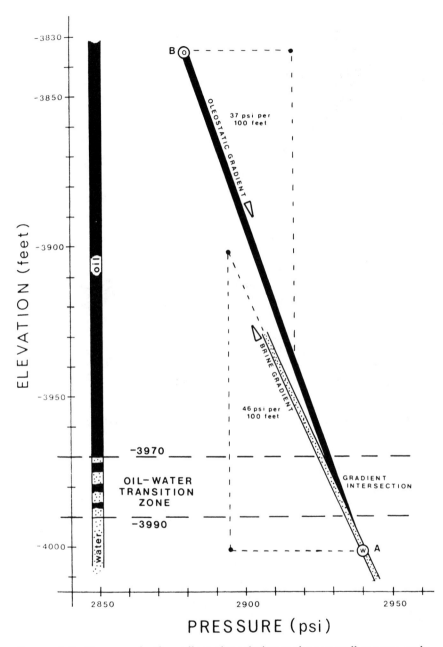

FIGURE A.6. Pressure–depth gradient plot solution to the two-well acreage evaluation problem in Figure 7.16. The brine gradient is projected upward from point A and the oleostatic gradient downward from point B. Since the gradient intersection occurs between −3970 and −3990 ft, all of the land parcels constitute viable oil prospects.

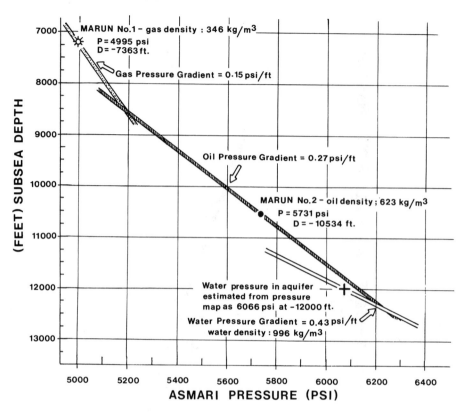

FIGURE A.7. Solution to pressure–depth gradient problem of Figures 7.17 and 7.18.

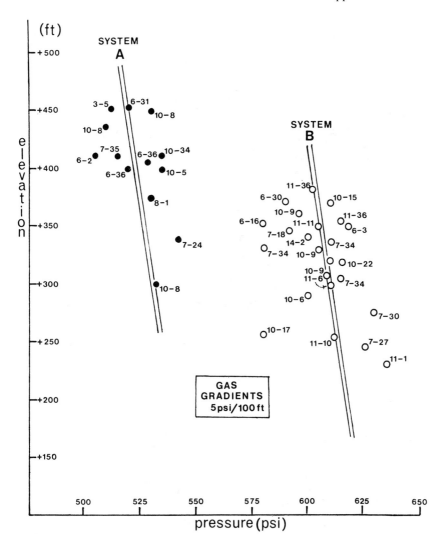

FIGURE A.8. Pressure–depth gradient plot for the Cretaceous channel data in Table 7.1, suggesting two independent pressure systems.

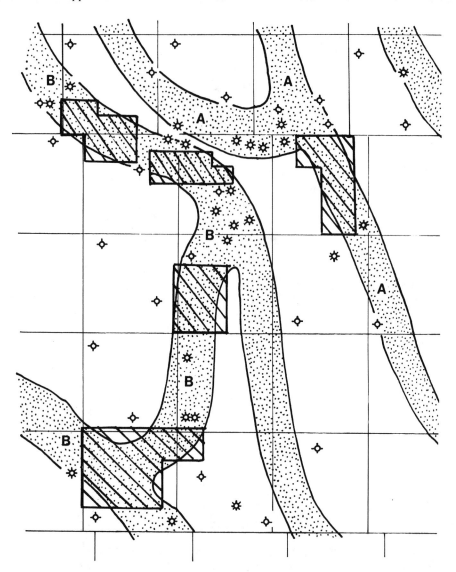

Figure A.9. The two systems defined on the pressure–depth plot, mapped in Figure 7.23, to define the channel configuration. Hatched blocks indicate potentially productive acreage for future drilling.

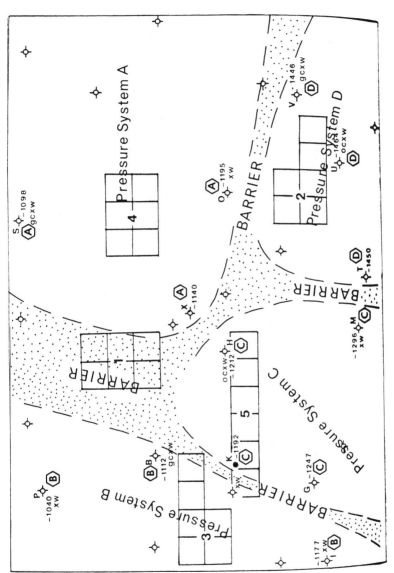

FIGURE A.10. Geographical locations of the pressure systems in the region shown in Figure 7.33, defined in the pressure–depth gradient plot. Land parcels "2" and "5" look desirable as possible prospects, since they lie in "pockets" downdip of implied permeability barriers.

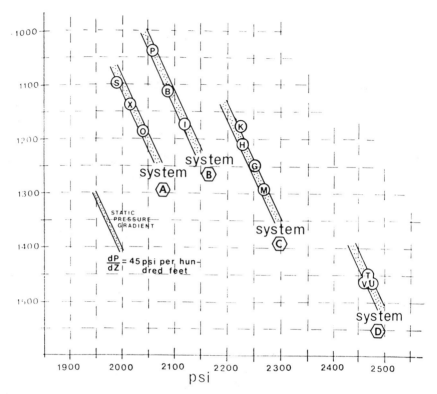

FIGURE A.11. Pressure–depth gradient plot for the data for the wells in Figure 7.34. The possibility of four separate systems separated by subsurface barriers is implied in this configuration, as described in the text.

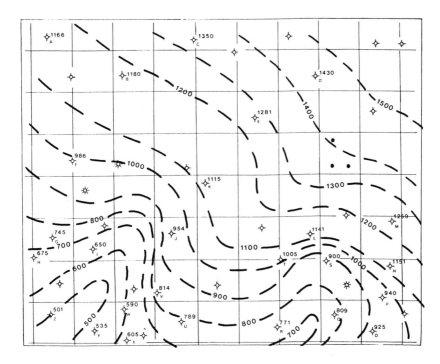

FIGURE A.12. Contoured regional potentiometric surface for area shown in Figure 8.31. The hydraulic head values are computed from the water pressure data of Table 8.1. Irregularities reflect locations of possible permeability barriers and potential hydrocarbon traps.

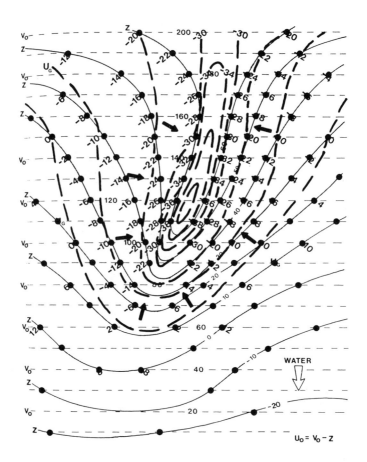

FIGURE A.13. Gas entrapment potential map showing the possibility of a gas accumulation that would result from the interacting structural and hydrodynamic conditions represented in Figure 11.14. Gas would migrate and be held in the closed "low" between the downdip-moving bottom water and the top of the structure. Under less extreme flow conditions, the structure would not trap gas, which would escape updip.

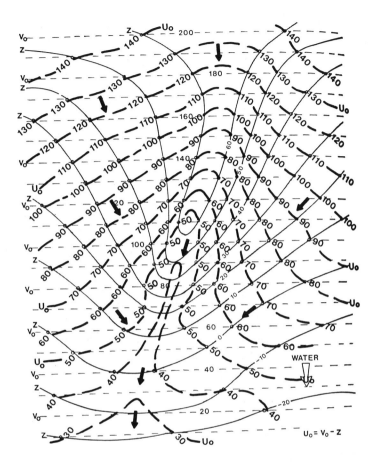

FIGURE A.14. Entrapment potential map for oil under the structural and hydrodynamic conditions represented in Figure 11.14. The construction discloses that there would be no oil entrapment under such extreme water flow; the structure would be flushed by the moving water. Arrows define oil migration paths.

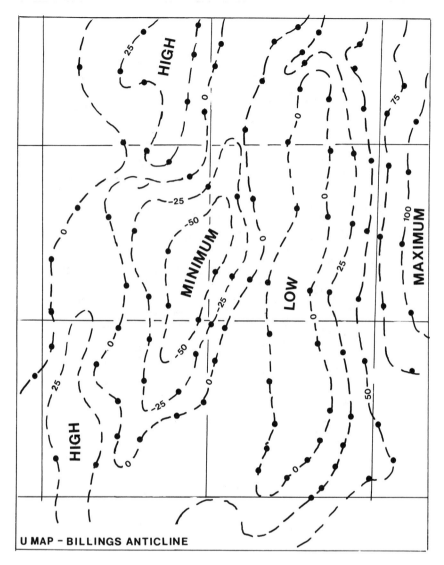

FIGURE A.15. Solution to Billings anticline oil entrapment mapping problem presented in Figure 11.17. This map shows the levels of potential for any oil dispersed through the unit. Highest closing contour defines size of the largest possible oil accumulation, in this case the "0" contour.

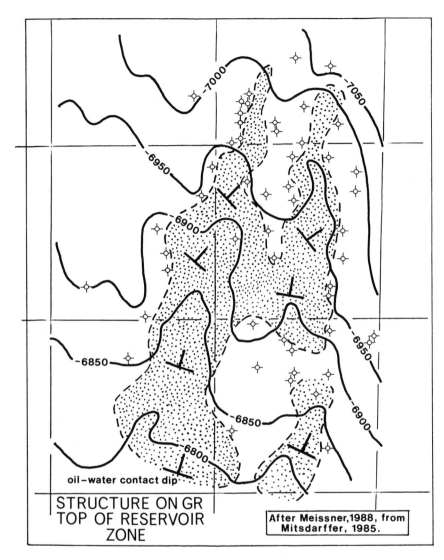

FIGURE A.16. Pool map of the Billings Anticline oil pool, to be compared with the locations predicted by the oil entrapment potential map.

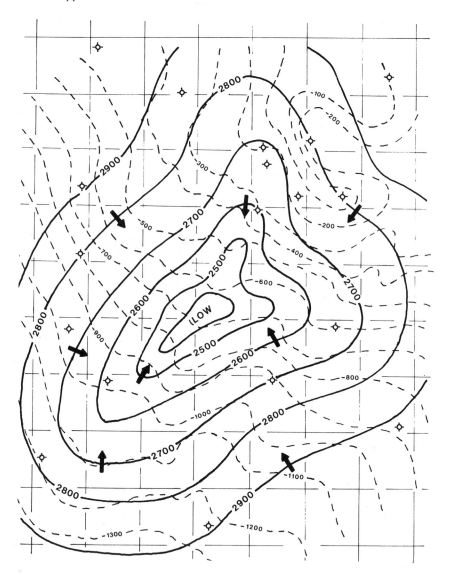

FIGURE A.17. Oil entrapment potential map constructed from the Devonian carbonate reef data in Figures 11.18 and 11.19 using the U, V, Z method.

Appendix B

Picking Hydrocarbon – Water Interfaces from Well Logs

Donald J. Fraser*

In the absence of accurate pressure data, the hydrocarbon/water interface may be determined from analysis of petrophysical well logs. Some careful analysis may be required to pick the true interface—the plane of zero capillary pressure.

Some terms are worth discussing:

Irreducible Water Saturation: This is the absolute minimum water saturation in a water-wet but hydrocarbon bearing zone. Water-free production should be obtained from this zone. (Think of the reservoir rock grains being water wet by a thin film of water; the remainder of the pore space is filled with hydrocarbon).

Transition Zone: Generally this should be avoided in well completion, as water may be produced.

True Oil–Water or Gas–Water Contact: This is the base of the transition zone, "the plane of zero capillary pressure." This is an important datum, as it may indicate the spill-point of a structural trap and is therefore related to the structural closure. (It is hopefully an indication of the volume of the reserves.)

All subsurface geologists and engineers should be aware of the classical indication of a hydrocarbon–water interface on well logs: the high resistivity in the hydrocarbon zone and the contrasting low resistivity of the 100% water-saturated zone. Often the hydrocarbon–water contact is not that obvious. In carbonates of variable porosity or in very high porosity sandstones, the resistivity contrast between the hydrocarbon-bearing versus water-bearing zone may be difficult to resolve. Two techniques are offered that may be employed to identify the true hydrocarbon/water interface:

---

*P. Eng. Fraser, Well Log Appraiser, Calgary, Alberta, Canada.

1. Resistivity Ratio Method of Hilchie.
2. Bulk Volume Water (BVW) Values.

Resistivity Ratio Method (Hilchie's Method)

This is a simple technique based on the ratio of the invaded zone (shallow reading) resistivity value (Ri) and the true resistivity (deeper reading) or Rt value.

This technique will only work in a zone that has been invaded by drilling mud filtrate. Invasion normally occurs in permeable rock after it has been exposed to the drill bit for a few hours. A test for effective invasion would be the presence of slight separation between the medium induction log curve (ILm) and the deep induction log curve (ILd) of the commonly used dual induction laterolog (DILL). The true resistivity value Rt is given by the deep induction log value ILd.

The invaded zone resistivity Ri is indicated by the so-called shallow reading Rxo tool—the Laterolog 8 device (LL8) in older dual induction laterologs (DILL) or the spherically-focused log (SFL) in more modern dual induction laterologs.

$$Ri/Rt = \text{constant in 100\% water saturated rock}$$

In the hydrocarbon bearing zone the Ri/Rt value is less. The water saturation (Sw) can be easily calculated by comparing Ri/Rt in the water zone to Ri/Rt in the hydrocarbon zone:

$$Sw = \frac{Ri/Rt \text{ in hydrocarbon zone}}{Ri/Rt \text{ in 100\% water zone}}$$

where Sw is the fraction of the pore space filled with connate water.

Dual induction logs are recorded on a logarithmic scale. Consequently, the ratios can be easily measured by the curve separations directly on the log. Simply measure the separation between the SFL curve (Ri) and the ILd curve (Rt) on a piece of paper. Place the paper on the 1 to 10 scale of the log with one "tick mark" on the 1 scale and read the ratio (separation) directly.

To pick the hydrocarbon/water interface move up and down the log looking for separation between the SFL curve (LL8 in older DILL logs) and the ILd curve. (This must be done in porous zones using SP, Gamma, and porosity logs as a guide). Then move up the log noting where the two curves start to converge (lower Ri/Rt ratio). This is the oil/water or gas/water interface. "Ticking" off the separation on a piece of paper, then moving the paper up and down the log makes this easier. The separation can be converted to a direct Ri/Rt value by overlaying on the 1 to 10 scale. (See examples). This technique works regardless of porosity, rock

Gamma Ray   Resistivity

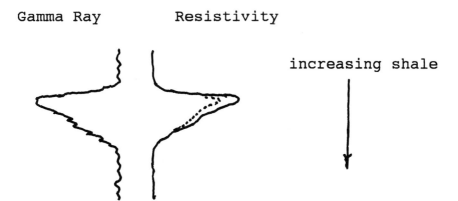

increasing shale

FIGURE B.1. Effects of shaly zones on the resistivity ratio.

type, or pore geometry, as long as the beds are permeable (and therefore invaded to some degree by mud filtrate).

Beware of the following hazards:

1. Tight Zones. In tight or nonporous zones the ratios close. (Therefore, check porosity logs).
2. Shaly Zones. Presence of argillaceous material causes the ratios to close, as shown in Figure B.1. Therefore use this technique only in clean, relatively shale-free rocks. (Normally the gamma ray log will distinguish clean from shaly rocks).
3. The lack of constant water salinity or water resistivity (Rw) over the interval of study. This is normally not a problem.
4. Pyrite. Because pyrite is conductive, it affects the conductivity curve and therefore increases the separation.

This technique can be used on other resistivity logs, but works best on modern tools. The older 16″ Normal-Induction log is recorded on a linear scale and requires calculation of Ri/Rt rather than direct measurement of curve separation. The Dual Laterolog (DLL) requires an invasion correction on the deep laterolog curve value (LLd).

Bulk Volume Water (BVW) ("c" Factor or Buckles Number)

The foot-by-foot computer calculation of bulk volume water (BVW) is another technique for identifying: the zone of irreducible saturation; the transition zone; and the hydrocarbon–water interface at the base of the transition zone.

$$BVW = \phi \times Sw$$

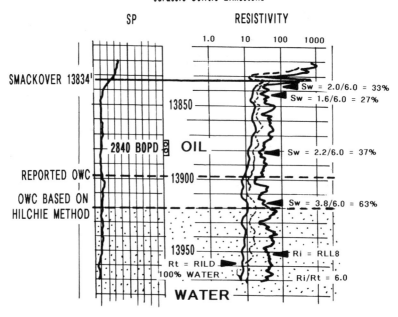

INEXCO #1 MASONITE

Jurassic Oolitic Limestone

FIGURE B-2.

(1) Inexco #1 Masonite, Prairie Branch Field, Clarke County, Mississippi. (Location N/2 SW/4 Sec 15-1N-14E, Clarke Co.)
Smackover Carbonate (Jurassic Oolitic Limestone).
Porosity 13834 to 14000'. (13517 to 13683 subsea elevation).

Depth	Ri/Rt	Ri/Rt (100% water)	Sw (Water Sat'n)
13836 Ft.	2.0	–	2/6.0 = 0.33 (33%)
13842	1.6	–	1.6/6.0 = 0.27 (27%)
13882	2.2	–	2.2/6.0 = 0.37 (37%)
13916	3.8	–	3.8/6.0 = 0.63 (63%)
13920 to 13990	6.0	6.0	6.0/6.0 = 1.00 (100%)

The oil/water contact is at 13920 based on the Hilchie method. The reported O/W contact at 13898 is still in the transition zone. The well was perforated 13874–13884 and flowed 2840 barrels oil/day through a 1" choke.

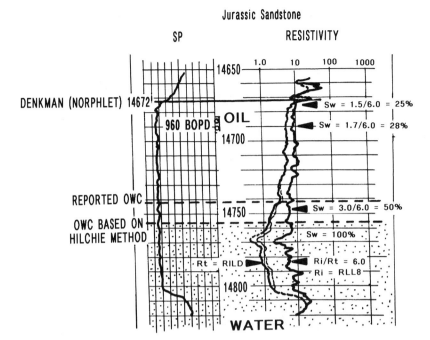

FIGURE B-3.

Denkman or Norphlet (Jurassic Sandstone). Porosity from 14672 to 14804 (same well as above).

Depth	Ri/Rt	Ri/Rt (100% water)	Sw (Water Sat'n)
14675	1.5	–	1.5/6.0 = 0.25 (25%)
14690	1.7	–	1.6/6.0 = 0.28 (28%)
14747	3.0	–	3.0/6.0 = 0.50 (50%)
14756 to 14804	6.0	6.0	6.0/6.0 = 1.00 (100%)

The oil/water contact is at 14756 (Hilchie method). The reported O/W interface at 14742 is probably the top of the transition zone. The well was perforated 14684–14694 and flowed 960 barrels oil/day through a 19/64" choke.

where:

$$\phi = \text{Porosity (fractional form)}$$
$$Sw = \text{Water Saturation (fractional)}$$

The BVW is the actual volume of the rock filled with water. For example a dolomite has a porosity of 10% ($\phi = 0.10$) and a water saturation of 30% (Sw = 0.30).

$$BVW = \phi.Sw = 0.10 \times 0.30 = 0.030$$

In this case the actual volume of the rock filled with water is 0.030 expressed fractionally, or 3%. (The remaining pore volume will be filled with hydrocarbons).

In brief, the BVW is at a constant value in the zone of irreducible water saturation (the oil log) for a given grain size and pore type. (The calculated BVW is for this reason used to determine that one is in the zone of irreducible water saturation).

Theoretically BVW equals ϕ in the zone of 100% water saturation (Sw = 1.0). In a computer well log analysis printout the BVW will therefore show with increasing depth: the zone of irreducible water saturation; the transition zone; and the hydrocarbon–water interface at the top of the 100% water-bearing zone.

This is a less sensitive and more complicated technique than the simpler Ri/Rt resistivity ratio method. (See for examples, Figures B.2 and B.3.)

Bibliography

Asquith, G.B., and C.R. Gibson. 1982. Basic well log analysis for geologists. *AAPG Methods in Exploration*, No. 3.

Fertl, W.H., and W.C. Vercellino. 1978. Predict water cut from well logs. In *Practical Log Analysis*. Part 4: *Oil and Gas Journal*, May 15, 1978–Sept. 19, 1979.

Hilchie, D.W. 1984. Water saturations determined by the ratio method: Talk given to the Canadian Well Logging Society. Calgary, Alberta, Jan. 19, 1984.

Hilchie, D.W. 1982. *Advanced Well Log Interpretations*. Golden, Colorado: Douglas W. Hilchie, Inc.

Appendix C

Relationship Between Pressure and Hydraulic Head

As often emphasized throughout the text, fluid pressure and hydraulic head (fluid potential), although related, are not the same. Although, in a static reservoir within the interstitial formation water, the pressures differ, hydraulic head is constant throughout the entire body of fluid. This is illustrated algebraically below.

The cross section in Figure C.1 represents a water-filled subsurface reef reservoir in which the pressure has been measured at two locations each at a different depth. These are P_A and P_B, and they are different from each other. The depths of the two recorders are rd_A and rd_B.

Therefore,

$$P_A \neq P_B \quad \text{and} \quad rd_A > rd_B.$$

P_A will exceed P_B, due to the greater weight of the overlying water, by an amount equal to the hydrostatic pressure gradient (grad P) multiplied by the difference in elevation between the two depths (corrected to sea level as shown in the figure).

$$P_A = P_B + \text{grad P} \, [(kb - rd)_B - (kb - rd)_A] \tag{C-1}$$

and

$$H_{W_A} = (kb - rd)_A + (P_A/\text{grad P}). \tag{C-2}$$

Substituting (C-1) in (C-2) for P_A, the equation becomes

$$H_{W_A} = (kb - rd)_A + \{P_B + \text{grad P}[(kb - rd)_B - (kb - rd)_A]\}/\text{grad P}. \tag{C-3}$$

This gives

$$H_{W_A} = (kb - rd)_A + P_B/\text{grad P} + (kb - rd)_B - (kb - rd)_A, \tag{C-4}$$

cancelling positive and negative terms yields. $H_{W_A} = (kb - rd)_B + P_B/\text{grad P}$ which equals H_{W_B}.

Therefore, under these conditions, $H_{W_A} = H_{W_B}$, and $H_{W_A} - H_{W_B} = 0$, even though $P_A \neq P_B$.

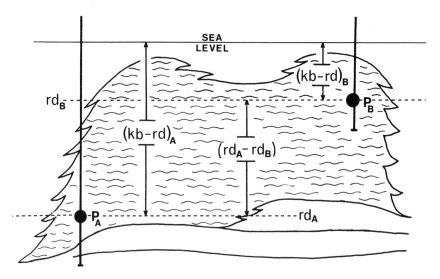

FIGURE C.1. Cross section showing the relationship between pressure and hydraulic head.

The two hydraulic head elevations H_{W_A} and H_{W_B} are demonstrated in this treatment as being equal, so that although the pressures in the reservoir vary between the two locations, the hydraulic heads are the same. The internal fluid potential is therefore constant.

Index